# Donkeys

# Donkeys
## Small-Scale Donkey Keeping

by Anita Gallion

HOBBY
H/F
FARM PRESS®

An Imprint of BowTie Press®
A Division of BowTie, Inc.
Irvine, California

June Kikuchi, Andrew DePrisco, *Editorial Directors*
Jarelle S. Stein, *Editor*
Jennifer Calvert, Lindsay Hanks, *Associate Editors*
Elizabeth L. Spurbeck, *Assistant Editor*
Cindy Kassebaum, *Art Director*
Karen Julian, *Publishing Coordinator*
Tracy Burns, Jessica Jaensch, *Production Coordinators*
Melody Englund, *Indexer*

Library of Congress Cataloging-in-Publication Data

Gallion, Anita, 1963–
  Donkeys : small-scale donkey keeping / by Anita Gallion.
    p. cm. -- (Hobby farms)
  Includes bibliographical references and index.
  ISBN 978-1-933958-95-8 (alk. paper)
  1. Donkeys. I. Title. II. Title: Small-scale donkey keeping. III. Series: Hobby farms.
  SF361.G35 2010
  636.1'82--dc22
                          2010005543

BowTie Press®
A Division of BowTie, Inc.
3 Burroughs
Irvine, California 92618

Printed and bound in China
14 13 12 11 10    1 2 3 4 5 6 7 8 9 10

*Dedicated to the wonderful,*
*supportive friends, family, and editors*
*who believed I could write this book*

# CONTENTS

# Why Donkeys?

In childhood, the fiery steeds of book and screen thrilled and enraptured us. We spent many an afternoon galloping around the neighborhood reenacting the Black Stallion's latest triumph over Flame or the amazing Seabiscuit versus War Admiral race. Horses represented strength, speed, and excitement, and we admired their wild, spirited natures. As the years passed, though, our taste in things equine began evolving. Those of us who had fulfilled our life-long dreams to own and ride our own horses realized that bones didn't bounce like they used to when unexpectedly ejected from the saddle and speed was more frightening than exhilarating when it came to thundering cross-country.

We began a fuzzy reexamination of our lives with equines, and when the image came into focus, the intriguing animal looking back at us reflected a somewhat mirrored image of our older selves—a little slower, a little plumper, yet certainly a little wiser. The donkey may be considered the wallflower version of the horse, but its more measured pace and laid-back personality have great appeal for older riders. Not that donkeys are an "old person's" equine. Their gentle, calm, friendly dispositions make them ideal for families with children, folks just starting with equines, and people with special needs, as well.

Donkeys have a plethora of uses. They can be fine pets, companions to other animals, and guardians for sheep, cattle, and goats. Manufacturers use donkey milk in soap and cosmetics; it's also sold for human consumption (more in other countries than in the United States). Raising and selling breeding stock can be a lucrative business. Recreational uses for donkeys include showing, packing, and burro racing. Donkeys are wonderful driving and riding animals; in fact, demand for trained Mammoth donkey saddle animals far outstrips supply. Whatever your interest, the humble donkey is waiting to serve you.

Not everyone appreciates the donkey's beauty, but for those who really look, it radiates from within. I cherish my association with these dignified, kind members of the Equidae family and hope to show you why donkeys can be ideal for your hobby farm.

# chapter ONE

# Donkeys 101

**W**aaaay back, 5,000 to 6,000 years ago, one line of prehistoric horses began to change. Their ears became longer and thicker; their manes took on a short, stiff, upright appearance; their tails formed a brush at the end; their neighs became brays; and their gestation lengthened from eleven to twelve months. *Equus asinus*, commonly known as the donkey, emerged.

Understanding the history of the donkey as well as acquiring an overview of donkey types, physical characteristics, social traits, and breeding will provide you with valuable information when you're ready to select donkeys for your farm.

## Donkeys through Time

Donkeys have been part of religion, myth, fable, folklore, literature, and proverb for so long that it feels as if they've been with us forever. Their story begins in parts of what is now northern Africa with the progenitor of the modern donkey, the African wild ass (*E. africanus*). With mitochondrial DNA testing, researchers have recently concluded that modern donkeys are descended from two subspecies of the African wild ass: the Somali wild ass and the Nubian wild ass. Both are currently considered critically endangered, with only a few hundred existing in the wild.

The donkey has never hurried things, and this personality trait has apparently been present from the beginning. Recent DNA studies have reached the conclusion that the domestication of the donkey was a lengthy process. Pinpointing a definite time frame for the domestication has been challenging for researchers because wild asses and their newly domesticated brethren bore a close resemblance. Scientists have had trouble differentiating between the two. During the domestication process, there was an overlap of time during which the donkey was both a hunted prey animal and a beast of burden.

*In late-nineteenth-century Central Asia, a cane vendor transports a huge bundle of sticks by donkey. Donkeys enabled people of Asia and Africa to increase overland trade and still play an important role there today.*

Archaeological evidence shows that the nobility of ancient Egypt hunted donkeys for sport, for hides, and for use in traditional medicines. Archaeologists have also discovered articulated donkey skeletons buried in special tombs within the cemeteries of several predynastic and early dynastic Egyptian sites, including Abydos (ca. 3000 BC) and Tarkhan (ca. 2850 BC). A 2008 study on the Abydos skeletons revealed signs of wear and strain on the animals' vertebral bones, an indication that people of the time used donkeys for riding and for carrying goods.

From the body morphology, scientists have determined that in terms of evolution, the Abydos donkeys lie somewhere between the wild ass and the modern donkey. Based on these findings and others, researchers have concluded that donkey domestication was still an ongoing process around 3000 BC.

The burial of donkeys in a high-status area indicates their value to the ancient Egyptians and the possibility that the royal household used these animals.

As the domestication process evolved, donkeys became tremendously important. With the help of friend donkey, people could travel more widely, which led to a large-scale redistribution of food and goods. This period saw a major shift in society, from an agrarian one tied to the land to a more mobile and trade-oriented one. The use of donkeys greatly expanded overland trade in Asia and in Africa. Today, in the arid, rugged regions where donkeys originated, people still use them to carry goods and provide transportation.

Many, many centuries after donkeys began making their way across Asia (and later Europe), they finally set hoof on the soil of North and South America. We have the conquistadors to thank for introducing

*Equus asinus* to the New World. The first asses—four jacks and two jennets—crossed the ocean in 1495 on a Spanish supply ship for Christopher Columbus. Subsequent ships brought horses as well as donkeys; both species eventually spread out across the Americas.

In the late 1700s, George Washington received two large jacks as gifts, both from European races chosen and bred for immense size and bone. These would be the forerunners of new a breed developed in the United States, the American Mammoth Jackstock. Breeders crossed the males of this species with mares to create the mule, a sterile hybrid work animal.

The king of Spain gave Washington the first jack, a large, ungainly, and unattractive Andalusian donkey named Royal Gift. Two jennets of the breed accompanied him. Washington's friend the Marquis de Lafayette sent the president the second jack, a Maltese dubbed Knight of Malta. This jack, which arrived with two Maltese jennets, was smaller but more vigorous than Royal Gift. Washington crossed an Andalusian jennet with the Maltese jack, a union that produced a large jack named Compound. With this jack, George Washington produced many valuable and much-sought-after mules. Thus the Father of Our Country unofficially became the Father of the American Jackstock and Mules, as well.

Meanwhile, the small, scruffy descendents of those asses brought over by the conquistadors were making themselves right at home in what would later be the American Southwest. Valued, as

*A prospector in the Old West uses a string of three burros to carry tools and other goods through the desert. Its ancestors' origins in the African desert made the burro (Spanish for "donkey") the perfect pack animal in the Southwest.*

they had been for centuries, for their hardiness and ability to survive in arid regions short on green grass and water, the donkeys of the Southwest (known there by the Spanish word *burros*) became favored by miners during the gold rush years of the mid-1800s. Adaptable and easygoing, donkeys lightened many a prospector's load, toiling as pack animals, pulling carts, carrying tools and supplies, and for the lucky, hauling gold.

Sadly, then as now, there was no such thing as job security. As mining became industrialized, miners cast the burros out to fend for themselves. They became the ancestors of the feral animals still roaming the region today.

By the early twentieth century, in America and other wealthy nations, the donkey's role had mostly changed from work animal to pet. In 1929, New York stockbroker Robert Green imported six Miniature jennies and a jack. A foal from these original imports was born on Columbus Day 1929 and, fittingly enough, received the moniker Christopher Columbus. He was the first Miniature donkey born in America.

Today, people use donkeys for riding, driving, guarding livestock against predators, siring mules, breaking calves to lead,

# Longear Lingo

Terminology connected to donkeys, mules, and the equine family in general can be confusing. Here's a little primer to get you started.

**Ass:** Derivative of the proper name of the species, *Equus asinus*; commonly referred to as a donkey or a burro
**Burro:** Spanish word for "donkey," used mostly to refer to the small feral animals of the southwest United States and Mexico
**Gelding:** A castrated male equine
**Hinny:** A (usually) sterile, hybrid equine, created by mating a male horse (stallion) with a female donkey (jennet). Looks similar to a mule and may require DNA testing to positively identify. The hinny is much rarer and more difficult to produce than the mule.
**Jack:** A male ass/donkey
**Jennet/jenny:** A female ass/donkey; jenny is slang for jennet
**Jenny jack:** A Mammoth jack used specifically to breed jennets
**John:** Slang term for a male mule, sometimes also referred to as a horse mule
**Longears:** Slang term for members of the ass family
**Mammoth donkey (also Mammoth Jackstock):** A breed of large ass. Males are 56 inches or taller, and females are 54 inches or taller.
**Miniature donkey:** Donkeys less than 36 inches in height
**Molly:** Slang term for female mule, sometimes also referred to as a mare mule
**Mule:** A (usually) sterile hybrid equine, created by mating a male donkey (jack) with a female horse (mare)
**Mule jack:** A jack primarily used to breed mares to make mules
**Standard donkey:** Donkeys from 36 to 54 inches

*"What'cha lookin' at Mom?" Mammoth foal Baby SweetPea pushes up beside her mother to check on going-ons outside the barn. Donkey foals are bright, curious, and anxious to be part of whatever is happening.*

and providing milk and meat. There are approximately 44 million donkeys worldwide, with the highest numbers in Mexico China, Ethiopia, and Pakistan, where most still work for a living. In America, the majority of donkeys are kept as pets.

## Donkeys at a Glance

Domestic donkeys belong to the Equidae family. The genus *Equus* includes asses, horses, and zebras. Donkeys compose the species *Equus asinus*. Let's take a closer look at how donkeys are classified and at some of their physical and breeding traits.

## Sizing Them Up

Unlike other species of livestock (cattle, goats, and horses, for instance), the number of actual donkey breeds is fairly limited. Instead, we primarily type and classify longears according to their size.

A few distinct breeds do still exist, and we will take a closer look at those breeds in the next chapter. For now we will examine the three main size classifications of donkeys.

## Mammoth Donkeys

Although size qualification varies among the official registries, in general jacks and geldings 56 inches and taller are considered

### DID YOU **KNOW**

Donkeys and other equines are measured at the highest point of their shoulders (withers), and the result is referred to in terms of hands. Four inches equals one hand. A donkey that is 60 inches at its withers is 15 hands tall.

*Feral donkeys roam freely around Custer State Park in South Dakota. Most feral donkeys, which live in loose-knit herds on federal lands, are standard donkeys.*

Mammoth size, as are jennets 54 inches or taller. Mammoths are the only donkey breed developed in this country.

In the late 1800s, Americans imported numerous jacks from large European breeds and crossed them with the country's smaller, native jennets. Continuing to select for great height and bone, breeders eventually developed the breed properly known as American Mammoth Jackstock. Mammoth Jackstock (or more commonly, if less properly, called Mammoth donkeys) are donkeydom's gentle giants. Originally developed to produce the draft mules needed before the advent of mechanized farming, they are now becoming increasingly popular for riding and driving as more and more people discover the quiet, well-tempered dispositions of these animals.

## Standard Donkeys

These are the midsize models. Standards range in height from 36 to 54 inches and can be subdivided further into small standard (36 to 48 inches) and large standard (48 to 54 inches). Most standard donkeys in America have descended from those feral denizens of the Southwest known as burros. The animals the U.S. Bureau of Land Management rounds up from federal lands and adopts out are standard donkeys. Many are large enough for a child or a small adult to ride, and they can be trained to pull a cart. Farmers and ranchers use this size donkey to protect livestock from predators and to break calves to lead. Pack burro racers most often use standard-size donkeys. Unlike Mammoths and Miniatures, standards are not considered a separate breed.

## Miniature Donkeys

Although the International Miniature Donkey Registry rules classify donkeys 38 inches or less as Miniatures (often referred to as Miniature Mediterranean donkeys, mini-donks, or minis), most people exclude donkeys over 36 inches. Minis make wonderful pets and can give toddler-size children a ride; they can pull an adult in a cart. Miniatures have become wildly popular over the last two decades and have been the most lucrative size donkey to breed and sell.

## Getting Physical

Though related to the horse and sharing some characteristics with this equine cousin, donkeys have their unique qualities. The most immediately noticeable difference is the long ears sported by *Equus asinus*. The other trademark feature of the donkey is its deep, raspy bray—a grating, rusty, sometimes squeaky "heeeee-haaaaw" sound, which is music to some listeners and fingernails on a chalkboard to others.

Donkeys are grazing herbivores with large, flat teeth well adapted for tearing and chewing plants. Primarily grass eaters, they also delight in nibbling small

bushes, bark, and limbs of trees, browsing somewhat as deer do. Like most grazing animals, donkeys grasp a plant with nimble lips, pull it into their mouths, and tear it off with their teeth. Because of

*Miniature donkey foal Chica seems to be giving her visitor a threatening, "back-off" look. Actually, she and mini companion Mona are being playful, a common characteristic of the smallest donkey type.*

# Why Keep Donkeys?

Our experts reveal the many reasons
why they chose to make donkeys a part of their lives.

### Chorus of Brays

"As for 'why donkeys,' for me there are many reasons. The obvious one is that they're so darn cute; but once you get past that, they're such forgiving, sweet-natured animals. For me, there's nothing quite like hearing that chorus of brays first thing in the morning when they hear the garage door go up and know that breakfast is coming. I can't help but smile when I hear that sound. Watching them run across the pasture with their heads stuck straight out and play is just too funny. I got my first donkey as a pasture mate to my horse, but I got so much more. Donkeys are just a delight, pure and simple."

—Lisa Lawhead

### Unexpected Sense of Humor

" 'Why donkeys?' Because they are intelligent, sensitive, and devoted, with a sense of humor one would not expect from an animal."

—Ann Firestone

### Mischievous and Special

" 'Why donkeys?' Because their personalities are so well suited to the needs of their people: friendly, loving, playful, sometimes mischievous, and willing to try anything they find reasonable. Because they read us so well and often reflect the attitude and mood of the people who spend

*I am shipping you to-day a pair of Burros Rocky Mountain Donkeys. Charges prepaid. It doesn't cost much to keep them as rage and tin cans are pie for them.*

*They are not expert of ears on autos. I hope they will reach their destination in good shape. F. R. H.*

time with them. Because they want to please and, with their 'can do' attitude, often surprise us with unexpected actions. Because they occasionally grace us with wonderful music—like angels from heaven. Because they fill a need, a hole in our souls and lives that we too often don't even realize exists.

"Because, as those of us who live and work with donkeys know, each one is special and better than all the others in some particular way. Yes, that's it. Why donkeys? Because donkeys are special!"

—Pat Liley

### Eager to Please

" 'Why donkeys?' Because they are nonthreatening. Because they are sweet, gentle, loving, and responsive, eager to do what you want—and they try very hard to understand what that is. With sensible kindness and enlightened training, they can do almost anything."

—Judy Merritt

*A rougher coat and longer ears, among other features, distinguish this standard donkey (left) from his fellow grazer, a horse. Standard donkeys make good companions for horses.*

their adaptation to marginal land, donkeys make efficient use of their feed and usually require less of it than do horses of comparable height and weight.

The donkey has a rougher, wirier coat than a horse does. Its mane is short, stiff, and normally upright, although in a few breeds (most notably Poitous and Poitou crosses) the mane can grow long and silky and fall over as a horse's does. The tail has short, wiry hair except for a "tuft" or "brush" near the tip. The donkey has a different pelvic structure than a horse has, the donkey's pelvis being higher, steeper, and narrower. This gives the donkey more of a pointed-hipped appearance.

Most donkeys have so-called white points. These include white rings around the eyes and muzzle, along with white on the belly and sometimes on the flanks and inner legs. Southern Mammoth breeders refer to a specimen with particularly vivid and extensive white points as "lit up," a much-desired trait. Many (but not all) donkeys have the strip of dark hair referred to as the cross. It travels from the mane all the way to the end of the tail (this part is known as a dorsal stripe) and has another strip of hair running across it over the withers (shoulders). The withers are flatter and less prominent than those of a horse, which leads to challenges for donkey owners in finding a saddle that fits properly. Donkeys do not have rear "chestnuts" (the small horny growths found on the inside of horses' legs). Their nasal passages are smaller than a horse's, which can make smaller nasal tubes necessary for veterinary procedures. In addition, donkeys have sixty-two chromosomes, as opposed to a horse's sixty-four and a mule's sixty-three.

## Socializing in a Bunch

Although donkeys are very social and much prefer to hang out with others of their kind, they live in the wild in loosely bunched groups, quite unlike the tight feral horse herds. In domestic herd situations, donkeys are also much less structured than horses are. The pecking order established among horses does not exist among donkeys. Jennies, for the most part, are placid and nonconfrontational. There will be an occasional pinning of

*A group of young Mammoth jacks gather at a fence. In general, donkeys don't mind crowding together and sharing their space; if there is a jenny around, however, jacks should not be housed together.*

# Biological Traits

**Temperature:** 99 degrees Fahrenheit
**Pulse:** 36–68 beats per minute
**Respiration:** 12–44 breaths per minute
**Expected life span:** 25–40 years
**Sexual maturity:** Jacks 8–18 months, and jennets 18–24 months. However, breeding is not recommended before the age of two for jacks and three or older for jennets.
**Heat cycle:** Jennets normally cycle every 21–24 days, remaining receptive for a period of 3–8 days, with ovulation occurring near the end of the receptive phase.
**Gestation:** 11–13 months, with 12 months the average.

the ears and threatening look at feeding time, but very little actual fighting occurs. Donkeys are generous with their personal space, allowing large numbers to crowd into a small shed or other shelter during bad weather.

Braying is common as a greeting (especially to the owner bearing breakfast). Jennies in estrus (heat) also bray, as do jacks when they are answering jennies or challenging other jacks.

Because they have a propensity for aggressive behavior, most domestic jacks are housed separately from other jacks and from jenny herds. Jacks can be very rough when meeting new jennets, and muzzling the jacks is a good idea when you introduce a jack and a jennet. Jacks may be kept together indefinitely as long as no jennies are within sight, smell, or hearing distance. Otherwise, jacks should be separated by two years old. Certain breeders do prefer

*A Mammoth jenny and her foal graze by themselves in a pasture. Initially, jennies tend to keep new foals away from the rest of the herd, which gives the newborn's eyesight and coordination time to develop so that the young foal can keep up with the herd.*

to keep their jacks running with a herd of jennies in a more natural manner and get along fine doing this. Owners should keep in mind, however, that some jacks attack and injure newborn foals.

Jennies seek solitude for their foaling. They also try to keep their newborns away from the herd initially, going so far as to stay out in cold, wet weather they would normally avoid. Because of these tendencies, most breeders try to keep an eagle eye on jennets due to foal, moving them to a clean, dry foaling area where the event can be monitored for the safety of all involved.

## Making Baby Donkeys

Donkeys generally become sexually mature between eighteen and twenty-four months, although maturity can occur much sooner, especially in the smaller donkeys. Because the possibility of an accident with these early bloomers always exists, experts recommend that you separate the sexes at weaning age.

Most breeders do not recommend breeding a jennet for the first time until she is at least three years old; four would be even better in terms of allowing her to achieve her full growth. Most jacks are fertile as two-year-olds and may start their breeding careers at a younger age than the jennets do.

Jennets normally cycle every twenty-one to twenty-four days. Many cycle year-round, although some stop during the coldest days of winter and the hottest of summer. Most remain in heat for three to eight days, with ovulation occurring near the end of the cycle. If bred and impregnated, they gestate the foal for approximately twelve months. Twinning is much more common in donkeys than in other equines.

# chapter TWO

# All Donkeys Great and Small

**U**nlike other types of livestock, donkeys have undergone relatively few physical changes in the thousands of years since their domestication. Although physical variations between donkeys have evolved—because of differences in use, care, type of forage available, and other environmental factors—true "breeds" of donkeys have always been rare and not nearly as specifically defined as they are in other domesticated species.

As mentioned in chapter 1, in North America, donkeys have mainly been categorized by size (and sometimes by color). The American Donkey and Mule Society has recently granted breed status to Miniature donkeys. Among most breeders and owners, it is also accepted that Mammoth donkeys (American Mammoth Jackstock) form a distinct breed, as well. For purposes of discussion in this chapter, we will examine the types or "breeds" of American donkeys as grouped by size.

## The Enchanting Minis

Without a doubt, Miniature donkeys are some of the most enchanting, affectionate, and lovable animals on Earth. The fuzzy, sturdy baby mini resembles a stuffed toy and is of a similar size. As adults, Miniature donkeys look like diminutive versions of the larger donkeys, averaging 32 to 34 inches at the withers, with a maximum of 36 inches allowed by the American Donkey and Mule Society. The International Miniature Donkey Registry separates minis into two classes: Class A, for 36 inches and under; and Class B, for 36.01- to 38-inch donkeys. Ideally, a Miniature will be compact and well-proportioned, weighing approximately 200 to 400 pounds. Miniatures come in a range of colors, the most common being slate gray or gray dun (one of many shades of grayish tan). Most Miniatures have a dark dorsal stripe that runs down the back and a second strip of dark hair across the withers; together they

*Miniature donkeys Chica and Mona drowse in the afternoon sun. Miniatures have a stuffed-toy appeal that is unbeatable.*

form a cross. Many minis have stripes on their legs, as well. Other colors include spotted, white, sorrel, dark brown, and black. Occasionally, minis will have a white star on the forehead or will be a solid dark color all over, with no white.

Unlike other breeds of miniature animals, Miniature donkeys have not been bred down from larger stock. The

minis originated on the Mediterranean islands of Sardinia and Sicily. According to the Miniature Mediterranean Donkey Association, because these donkeys were so small, peasants would bring them into the house to toil at turning grinding stones. Some eighteenth-century woodblock pictures depict donkeys harnessed to a grain mill, blindfolded, and forced to walk in endless circles. In addition, the Miniature donkeys toted wood, water, and supplies to earn their keep. Unfortunately, crossbreeding with larger donkeys has made the original tiny breed almost extinct on the islands today.

As mentioned in chapter 1, in 1929, Robert Green, a wealthy stockbroker in New York, imported the first Miniature Mediterranean donkeys to the United States. Their number rapidly grew over the next three decades, and in 1958, Bea Langfield of Omaha, Nebraska, created the Miniature Donkey Registry (MDR).

Mrs. Langfield operated the MDR until 1987, at which time she turned the registry over to the American Donkey and Mule Society, which continues to maintain records and register Miniature donkeys, along with donkeys of other sizes, mules, and zebra hybrids.

Fortunately, in contrast to the hard life of its ancestors, today's Miniature donkey generally leads the life of a pet. Because of their engaging, companionable nature and small size, these easily kept creatures have become the most popular breed/type of donkey in America, a favorite of folks of all ages. Minis fascinate children, and seniors enjoy being able to keep and care for an equine whose small stature makes husbandry less physically challenging. These little donkeys can be kept on small acreage

*In this blurry photo from 1946, Robert Green affectionately greets a Miniature donkey, the breed of donkeys he established in America seventeen years before.*

*A herd of Miniature donkeys races across the pasture at Hester Farm in Dexter, Missouri. Miniatures are the most popular breed of donkey in the United States.*

# Einee, Meenie, Mini, Mammoth

Our experts explain what characteristics drew them to their particular breed or type of donkey.

### Just Like Potato Chips

"Miniature donkeys are just like potato chips: the more you have, the more you want! Six years ago, as we were planning for our retirement, we decided to sell the home that we raised our children in because, as empty nesters, we wouldn't need that much space to clean, air-condition, and heat. At the time, the AARP buzzword was *downsizing*! So while we were downsizing our home, we also downsized our equine: we sold our two horses, bought two bred Miniature jennies, and the rest is history! We now have a herd of about fifty Miniature donkeys and cannot imagine spending our retirement years doing anything else but giving daily donkey hugs and butt scratches!"

—Harvey and JoAnn Jordan

### Remembrance of Childhood

"As for what makes standard donkeys special for me, I guess it's because that is what I grew up with here in the UK. You never saw anything but a standard. We got used to seeing them on the beach giving children rides, and a lot of children had them as a first equine before they moved on to a pony. I actually had an allergy to all equines as a child, so could never even learn to ride. It wasn't until my twenties that I discovered I could venture closer and not sneeze for hours on end! So I bought a pony. But I always wanted to have a donkey,

and it seemed like a good idea to get one as a companion. When I finally had my own place with some land and the pony came back from livery, we saw two donkeys for sale, and I went to see them. From the second I saw them at the dealer's, I was smitten. So Hector and Smokey came to live with us. Their big brown eyes, their incredible personalities, their intelligence, and the love they gave us hooked me in instantly. So, for me, it is the standard donk I adore!"

—Liz Barrett

### Big Puppy Dogs and Gentle Giants

I have been a dog person since childhood and a cat person for thirty-one years (after meeting my husband). We acquired our first equine just about a year ago, Ms. Luna Lovegood, the Haflinger molly. We had the chance to visit the farm where she was born and got to meet Jack, her dad. What a big puppy dog! I've been sold on the 'gentle giants' ever since and now have four Mammoth donkeys."

—JoAnnie Kale

### Calm, Smooth, and Trusted Mount

"I got sick and was diagnosed with lupus. I was riding a bombproof Arab mix. She was fun, but she'd still have 'horse' moments, and after I became ill, I started worrying more about an explosion and

not enjoying the ride. I bought a mule, which just wasn't a good match for me, and ended up wrestling the entire ride. I was just about to give up riding, when I remembered reading about Mammoth Jackstock. After a few months of research, I found a Mammoth jennet who had been ridden once and was four years old. We went and looked at her. The seller rode her, then Jon rode her, and we brought her home.

"I was a little nervous, as I wouldn't ride a horse that green. Jon rode Matilda about six times, and finally he said, 'You really need to try her.' I got on her, and I've never gone back! She is great, she's smooth, her 'spooks' are nothing. She takes care of me as her rider, and I've ridden trails with her that I wouldn't have trusted to ride on a horse."

—Tanya Tourjee

### Miniature Size, Big Heart

"Donkeys are donkeys. So why the little ones? Is it because it is natural for humans to adopt the smaller animal? Because they are easier to handle, eat less, take up less room, are easier to load and unload? All of that, yes, but mostly it's because when I kneel, their soft noses kiss softly with my nose, they trade breaths with my lungs, and those brown eyes meet mine. They aren't miniatures at all. They just made their world bigger."

—Russ Hauenstein

(zoning laws permitting) and have modest feed and housing needs (compared with those of a full-size equine).

Its docile disposition makes the mini a safe pet for people with disabilities. Miniatures also are great ambassadors when it comes to visiting nursing homes and performing in parades, petting zoos, and nativity scenes. In addition, Miniatures willingly give rides to small children and pull them in carts; minis can even pull an average-size adult with no problem. Hikers use Miniatures as pack animals. If the load is properly packed and well balanced, an average-size mini can carry 75 to 100 pounds.

Showing Miniature donkeys at mule and donkey shows is extremely popular, as well. Shows offer a vast array of classes, including ones on coon jumping, driving, and in-hand leading through trail obstacles. The shows also have halter classes, in which the donkeys are judged for proper conformation.

Breeding Miniature donkeys to sell has been a lucrative business for a number of people and a passionate hobby for many others. Given the downturn in the entire economy as well as in the equine industry itself, you should consider the costs and odds of turning a profit before starting a breeding program. There will always be demand and good prices for the highest-quality animals, but the market can quickly become flooded by a combination of overbreeding and a decline in demand caused by a bad economy.

If you want to establish a breeding program for Miniature donkeys (or for any other donkeys), do your homework first. Know your product and your market, and acquire the best breeding stock that you can. It's a competitive industry, but one that's financially rewarding for those who produce the best.

*A word of caution*: Do not use Miniature donkeys as guards for animals such as sheep, goats, and cattle. They are too small. In fact, there have been tragic consequences when the guardian mini has become the prey rather than the protector. The minis are simply no match for a pack of coyotes or marauding dogs. Standard-size donkeys serve this purpose much better.

## The Underappreciated Standards

You could say standard donkeys are the Rodney Dangerfields of the donkey set—they get no respect. They are the somewhat underappreciated members of an already underappreciated species, not as cute as the cuddly Miniatures nor as awesome as the imposing Mammoths. Most standards in America are descendents of donkeys released by or escaped from Spanish explorers, miners, ranchers, and Native Americans over the past several centuries. People frequently refer to standards as *burros* (Spanish for "donkeys").

When most people think of donkeys, the standard (36 to 54 inches) usually comes to mind. Standards are sometimes further categorized into small standard (36.01 to 48 inches) and large standard (48.01 to 54 inches). Occasionally, two Mammoth parents will produce an undersize offspring that will only mature to standard size, or two Miniatures will have an unexpected "Baby Huey."

Because standards are not specialized, as the Miniatures and Mammoths are, no specific registry dedicated to standardizing their appearance and characteristics exists. Many, such as the feral burros managed by the Bureau of Land Management, have evolved as a result of natural selection. Standard donkeys come in many colors, but as with minis, the most common is the slate gray or gray dun, with leg striping and the back and withers cross. Spots are common, as well, and very popular.

*At Custer State Park, where they freely roam and forage, feral burros stop to check out a visitor's car. Their height places them in the category of standard donkeys.*

*A spotted standard at Custer State Park gingerly accepts a carrot from a small tourist. The U.S. Bureau of Land Management makes some wild donkeys available for adoption.*

In general, standards do not have as much financial value as their smaller and larger brethren do, and they will be the cheapest of the three sizes to purchase. Yet standard donkeys unquestionably have worth. The midsize donkeys are well suited for a myriad of jobs. Many ranchers and farmers have taken advantage of donkeys' inherent dislike of coyotes and dogs by putting a longear or two out to protect the cattle, sheep, and goats by chasing those predators from the pasture. Standards have proven to be the best size donkeys to utilize as livestock guardians. Jennies and geldings are much better suited to the task of guarding than are jacks, which tend to be too aggressive and may injure smaller

livestock. Standard donkeys are also the right size and disposition to use as companion animals for high-strung racehorses and newly weaned foals.

Standard donkeys can certainly be handy helpers around the homestead, as well. You can hook a fairly simple sled or travois to your donkey to skid firewood, clear brush, or haul trash. You can also employ these donkeys to break calves or colts to lead. To do this, attach a collar on the donkey by a short lead to the halter of the "student," in a small, supervised enclosure. A battle of wills commences at this point—one the donkey always wins.

Large standard jacks are very popular among saddle-mule breeders as some

*Giant Andalusian donkey Connan (left) and elderly companion Woody (thought to be of Andalusian descent) greet children visiting the family-run Nerja Donkey Sanctuary in Spain.*

## European Breeds That Helped Create the American Mammoth Jackstock

| BREED | SIZE | COLOR |
|-------|------|-------|
| Andalusian | 14.2–15 hands (58–60 inches) | Gray |
| Catalonian | 14.2–16 hands (58–64 inches) | Black, brown |
| Majorcan | 15.2 hands (62 inches) | Black |
| Maltese | 14–14.2 hands (56–58 inches) | Black, brown, blue (gray dun) |
| Poitou | 13.2–15 hands (54–60 inches) | Dark brown, black |

feel that the bigger, draft-type Mammoth jacks produce too coarse an offspring. The breeders prefer a finer-boned, more athletic standard jack for this purpose. The larger standards are also quite usable for riding and driving, especially for children and small adults.

## The Disappearing Mammoths

As discussed previously, the American Mammoth Jackstock (also referred to as American Jacks and Jennets, American Standard Jacks, Mammoth donkeys, and Mammoths) is a unique breed of large ass developed in this country from imported European breeds mixed with our native stock. The need for a better donkey to sire a superior type of draft mule was a driving force for developing the Mammoth. In late-eighteen-century America, the few mules to be found were sorry specimens that had been imported from the West Indies by New England farmers. Intensely interested in improving the selection, George Washington started with the jackstock given to him by the king of Spain and the Marquis de Lafayette, and he continued to raise both fine jackstock and the mules they produced at his Mount Vernon estate until his death. Another early pioneer in jackstock breeding, Kentucky statesman Henry Clay, bred many donkeys and mules at his Ashland estate during the early nineteenth century.

During the mid- to late 1800s, Americans imported several thousand jacks from Europe. In fact, so many were taken from their native lands that it severely depleted the gene pools in those countries, leading to the extinction of several breeds. Most imported jacks were of the Andalusian, Catalonian, Majorcan, Maltese, or Poitou breeds (see breeds chart below). In the United States, these breeds

| ORIGIN | GOOD QUALITIES | LESSER QUALITIES |
| --- | --- | --- |
| Andalusia, Spain | Heavy bone | Sometimes thick jawed and Roman nosed |
| Catalonia, Spain | Fine-textured hair, good size, stylish | Occasionally light boned |
| Majorca, off the coast of Spain | Huge size | Sluggish disposition |
| Island of Malta in Mediterranean Sea | Very vigorous | Small size |
| France | Heavy bone, lots of substance | Rare, high priced |

*American Mammoth Rebel poses at a show in Tennessee. A mature Mammoth jack is a large and imposing specimen of a donkey.*

were crossed with each other, along with native stock descended from earlier Spanish imports, in an attempt to produce the large, heavy-boned, draft-type jack desired by the draft-mule breeders of the day. Tennessee, Kentucky, and Missouri, in particular, became known for their larger numbers and fine quality of jackstock. Breeding jackstock proved a high-dollar endeavor for most of the nineteenth century and the first half of the twentieth century.

Unfortunately, with the advent of the tractor and mechanized farming, the demand for mules and for jackstock plummeted. From the 1920s through the 1950s, numbers dropped alarmingly. The American Mammoth Jackstock is now listed as critical on the American Livestock Breeds Conservancy's Priority List. The critical category comprises breeds with fewer than 200 annual registrations in the United States and an estimated global population of fewer than 2,000.

Once merely a means to an end—making mules—in recent years Mammoths have been gaining favor for their potential as *using donkeys* (that is, donkeys people use for specific activities, such as riding, pleasure driving, and plowing gardens). With its quiet, calm disposition, the Mammoth appeals to people looking for a low-stress riding and driving mount.

Mammoths are easy to train, and their friendly personalities make them a pleasure to work with. These donkeys can be particularly good as riding choices for children, for handicapped individuals, for people who have confidence issues (possibly stemming from a horse-related

# What's in a Name?

What's in a name? A lot of differing opinions, apparently. The largest breed of American ass has been called many things since its beginnings some 200 years ago. In 1888, the original registry was dubbed the American Breeders Association of Jacks and Jennets. At this time, owners simply referred to the large asses as American jacks and jennets (*jackstock* being the plural term). In 1923, the registry merged with a competing association, the Standard Jack and Jennet Registry, and the latter name was kept. The word *standard* referred to its "standardized" breed characteristics, but this use of the term *standard* ultimately led to a certain amount of confusion because people also referred to smaller, burro-type donkeys as standard donkeys. So in 1988 the registry decided that it would change its name once again, this time to the American Mammoth Jackstock Registry. And so the word *mammoth* was thrown into the mix of terms. Purists insisted the proper term for the breed was now Mammoth Jackstock or American Mammoth Jackstock, but many old-timers never did pick up the term *mammoth,* and they continued to refer to these donkeys just as jacks and jennets. Other people who were unfamiliar with the term *jackstock* began referring to these asses as *Mammoth donkeys*. This seems to be the term that is most widely known and the one most frequently used by people today.

*A Mammoth donkey, Apollo, contemplates a much smaller, canine resident of the farm. Over the past 200 years, American Mammoth Jackstock have been known by many different terms.*

# The Catalonian Donkey

American Jackstock breeders have long prized Catalonian jacks and jennets. In *The Breeding and Rearing of Jacks, Jennets, and Mules* (1902), L. W. Knight remarks, "The Catalonian is a jack of great style and beauty and of superb action. Those who are engaged in rearing them need never fear but that the demand of them will be active and the prices remunerative." These donkeys, from northern Spain, composed most of the nineteenth-century European imports Americans used to found their Mammoth breed. In fact, Catalonians were in great demand worldwide for breeding jackstock; between 1880 and 1910, Spain exported more than 1,700. That took its toll on the Catalonian, whose numbers dropped dangerously. During the 1940s and 1950s, Spain's isolation and reliance on self-sustained, traditional farming led to a resurgence in the breed's numbers. They fell again with the advent of industrialization and mechanized farming. Today only 400 to 500 documented pure-bred *Catalans* (the Spanish term) exist worldwide. Jose Manuel Fernandez of Gerona, Spain, has one of the largest, most prestigious Catalan farms. He, the Yeguada Militar (Spanish military equine breeding program), a university restoration program, and two other breeders work together to restore biological viability to this magnificent breed. Mr. Fernandez, a third-generation Catalan breeder, has a herd with all four of the major bloodlines, including two derived from the Yeguada Militar's program. Many Yeguada Militar donkeys have pedigrees going back several centuries.

*Bred by Jose Manuel Fernandez of Gerona, Spain, this prize-winning Catalonian donkey named Ecijano displays the energy and striking appearance for which pure-bred Catalonian donkeys are famed worldwide.*

*Margarita, a Mammoth, totes one youngster and leads another across the river. Mammoth donkeys have become increasingly popular as riding animals.*

accident), and adults who want a donkey but are too big to ride standards.

Physically, Mammoth Jackstock are large, clean-boned, sturdy donkeys with exceptionally well-made heads and long, fine ears. They come in the same colors the smaller donkeys do, except that the gray dun with stripes is much less common. Type varies greatly, depending on the donkey's intended use. As discussed above, the American Jackstock were originally created as large, sturdy, heavy-boned, thick-bodied animals for the purpose of producing strapping draft mules. Today, preservationist breeders

continue to strive for that type, as do draft-mule breeders. The largest emerging market for Mammoth Jackstock is for riding/driving donkeys, as well as for sires for saddle mules. More refined Mammoth jacks are being used to produce the show-type mule and donkey, while the heavier boned are most often seen in draft-mule production.

There are two primary registries for the American Mammoth Jackstock. The American Mammoth Jackstock Registry (AMJR) requires that jacks be at least 58 inches tall and jennets at least 56 inches. In addition, both sexes must have at least

*The distinctive cadanette of shaggy, matted hair tells the observer that this is a rare Poitou donkey. This Poitou was photographed in the breed's native country, France.*

a 61-inch girth. Jacks must measure at least 8 inches around the cannon bone (measured just below the knee); jennets must measure 7½ inches. The American Donkey and Mule Society (ADMS) requires a height of at least 56 inches for jacks and geldings and 54 inches for jennets. Both registries will register on size alone, even without a known pedigree.

## Other Breeds in North America

There are two other donkey breeds found in North America: the Poitous and the spotted asses. Donkey breeds elsewhere in the world—including the Andalusian, Catalonian, Majorcan, and Maltese—are no longer found on this continent in any significant numbers.

## The Ancient Poitous

The Poitou (pronounced "Pwa-too") is an ancient breed from France, dating back nearly 2,000 years and possessing a breed registry some 300 years old. Although they are not exceedingly tall, the Poitous are powerfully built, with great bone and substance. Their most notable characteristic is their long, matted, "dreadlock" coat of dark brown or black. In their native France, the Poitous do not normally have their coats trimmed or groomed; in fact, breeders prize the shaggy, corded mess, called a cadanette. The Poitous, greatly valued as mule producers, have always commanded extremely high prices. Nowadays, there's another reason Poitous are expensive: their rarity. Only an estimated 400 pure and part-bred animals can be found in the world. However, preservationists in France and America are bent on saving this critically endangered breed through their breeding programs. Thanks to these efforts, Poitous, having teetered on the brink of extinction for decades, are slowly recovering.

## The Up-and-Coming Spotted Asses

The American Council of Spotted Asses (ACOSA) incorporated in 1969 with the intention of establishing a breed of spotted asses. Donkeys (along with mules and hinnies) are eligible for registration if they have two spots behind the throatlatch (the throat area of the donkey where head and neck meet) and above the legs. Over the past ten years or so, spotted asses have been become very popular and have brought good prices, both at public sales and in private transactions.

*These pretty spotted donkeys would be eligible for registration by the American Council of Spotted Asses, one of several donkey registries.*

# chapter THREE

# Choosing and Buying Your Donkeys

**B**uying donkeys is fun. It's more like adding members to the family than like buying livestock for the farm. That's not to say there aren't challenges associated with buying what's right for you and your situation. This is when doing your homework pays off. Know what you want, what it costs, and where to find it when you are ready to buy.

## Deciding on the Right Donkey

Are you looking for a Miniature, a standard, or a Mammoth? Perhaps you're drawn to the exotic Poitou (get out your checkbook!) or a flashy spotted ass. Do you want a pet or a performance animal, or are you thinking of establishing a breeding program? (Consult chapter 2 when deciding what breed or type is right for you.) Prices and availability—two aspects you should also take into account as you define your search criteria—will vary tremendously. Other factors to take into account are number of donkeys, sex, age, and level of training. Once you've seen the donkeys, ask yourself whether they are what they've been represented to be, whether they fit your needs, and whether they are animals you'd be proud to own. Somewhere out there await your perfect donkeys.

## Donkey or Donkeys?

Most donkey breeders will explain to you that you need at least two donkeys to meet the donkey's deep desire for the company of its own kind. A lone donkey is an unhappy donkey. Although other types of animals and livestock can be used for some companionship, for a longear nothing takes the place of another donkey. There's no need for you to worry that your donkeys will bond to each other rather than to you. In general, they will compete with each other for the petting, scratching,

*Eulalie, a Mammoth jenny, checks out a young visitor to her pasture. Social and curious by nature, donkeys enjoy interacting with people; in fact, they take great interest in everything and everyone around them.*

and all the other kinds of attention their owners will happily bestow upon them.

## Female or Male?

You will need to decide whether a male donkey or a female donkey (or donkeys) will suit your purposes. Unless you are serious about becoming a breeder and keeping a good-size herd of jennets, you should scratch jacks off your list right away. Jacks are *not* pets and have serious responsibility and liability issues associated with them. Even a 350-pound Miniature jack can prove to be a deadly

# Before You Shop

Before going in search of your donkeys, examine your lifestyle and facilities. Donkeys don't need extravagant barns and living quarters, but they do need, at minimum, a three-sided shelter of some type to keep them out of wet, windy conditions. Do you have safe and adequate fencing? Count on having irate neighbors and local motorists if you don't. You will also need to have access to a good equine veterinarian, a farrier, and a hay supplier if you don't grow your own hay. Last but not least, do you have time? Donkeys aren't terribly time consuming, but any type of equine should be checked on and fed at least twice daily.

and dangerous animal under certain circumstances. They are hormonal time bombs, and the gentlest jack can lose control in the blink of an eye when distracted by a jenny in heat or a perceived rival. Jennets and geldings are much, much better suited for anyone except those who require the jack for breeding.

There are a few differences, though not many, between jennies and geldings. Geldings seem a bit goofier and more playful than jennies, which are somewhat serious and staid in comparison. Because of her breeding potential, you will probably pay more for a jenny than for a gelding.

## Trained or Untrained?

If you are planning to have a donkey for riding, driving, or any other specific

activities, you will need to decide whether to shop for a trained or an untrained one. Donkeys are normally very easy to train. If you have any sort of equine experience, you will probably find that you can train one yourself.

Donkeys don't often react to being saddled or mounted, even the first time, and will quickly pick up the basics of saddle work when ridden with other calm, quiet animals. You'll find it is much

*At a show, the sign on the stall of this jack clearly warns passersby that he bites. Jacks do not make good pets and should only be kept for breeding purposes.*

*Donkey keeper Robert Auge demonstrates packing with a Mammoth gelding. Their calm and willing dispositions make donkeys ideal for this type of activity.*

# Auction Avoidance

Auctions are not a great place to buy donkeys. Although prices there are often low, most of the time there is a reason (and not a good one) that people have consigned those donkeys to a sale barn rather than selling them privately. Barren jennies and sterile jacks, not to mention foundered (lamed) or otherwise unhealthy stock, frequently end up at sales barns. Sometimes unscrupulous breeders or sellers will try to sell baby donkeys as "bottle babies" (orphans raised on bottles rather than on their dams), especially at "exotic" animal auctions. Baby donkeys should not be purchased until they are of weaning age, no younger than five or six months.

*A sweet-looking baby Mammoth rests comfortably in a pen at Coyote Lane farm. First-time owners should keep in mind that foals won't be ready for breeding, riding, or driving for three or four years.*

more difficult to locate saddle-trained donkeys (especially Mammoths) as the demand is far greater than the supply.

## Youngsters or Adults?

While it's fun starting with a youngster, sometimes waiting for it to be old enough to use in some capacity can feel like an eternity. Most experts recommend that jennies not be bred until three years old, and neither sex should be ridden before that age. However, basic training, such as teaching good leading and handling manners, as well as ground driving, can begin when the donkeys are yearlings. Donkeys can do some light driving at two, but it's best to wait until the age of three for anything more strenuous.

## Finding Your Donkeys

Where do you find your donkeys? One place you shouldn't try is an auction (see "Auction Avoidance," *opposite page*). Look for reputable breeders. One of the best ways to find breeders in your area, and to educate yourself on all things donkey, is to surf the Internet. Online

*Mounted on Mammoth donkeys, riders wait for instruction at the Houston Livestock Show and Rodeo. This type of venue is good for meeting breeders and learning about donkeys.*

you'll find a number of longear-specific classified ad sites (see this book's Resources) and general equine sale sites as well as all-purpose sites such as Craigslist. You'll also find ads in breed publications (listed in Resources).

Other ways to search out those perfect donkeys include joining a local mule and donkey club (members will assist with your search), attending shows, and becoming a member of the American Donkey and Mule Society. Its bimonthly magazine, *The Brayer*, is full of ads. You can ask the people at your local feedstore or veterinarian's if they know of any donkeys for sale. One of the most rewarding ways to add donkeys to your family is through adoption. There are several fabulous private, not-for-profit, longear rescues (see Resources); your support of them can give abused, neglected donkeys a new lease on life.

When you locate a breeder or other source, arrange a visit so you can see the stock up close and personal and ask questions about the breed and its care. Below are some tips for determining whether you have found the right donkeys.

## Size Up the Merchandise

Experienced owners will tell you in no uncertain terms that potential buyers should invest in a measuring stick. Many tack store and horse-supply catalogs will have the kind you need for measuring donkeys. If you want to buy donkeys for a breeding program, you will certainly care how tall they are. Inches can literally mean hundreds of dollars' difference in the value of Miniatures and Mammoths.

Likewise, if you are buying a donkey to use for riding, you want to make sure it's large enough. Not to make donkey sellers sound unscrupulous, but size is something that gets exaggerated (both ways) with alarming frequency. As one American Mammoth Jackstock breeder put it, "Watch out for the word *about* being used by a seller in regards to size. Often it means there is some variance from what's been quoted, and this can vary from fractions of an inch to fractions of a yard." Another breeder refers to his measuring stick as his "lie detector." It's just good insurance to take your stick and double-check the size.

## Look for Health Problems

When you arrive to examine your prospects, take a look at the surroundings. Most farms aren't palatial showplaces, but they should be safe and tidy environments. As you assess the stock, keep your eyes and your ears open for runny

# Buying Sight Unseen

The Internet makes it easy for us to buy "sight unseen" from hundreds, if not thousands, of miles away. This can be successful as long as the buyer does his or her homework on donkeys ahead of time. Ask the seller for some references, for a size guarantee, and for health and fertility (if applicable) guarantees, preferably in a written contract. Invest in a prepurchase examination from a veterinarian close to your prospect. The vet may be a good person to take a "second-opinion" measurement. When conducting a long-distance transaction, make sure everything is in writing to protect all parties.

*Using an equine measuring stick, Bobbi Ward checks the height of this Mammoth, Lurch, who stands 62 inches high (or, in equinespeak, 15.2 hands). Make sure the donkey is standing in a level area, then place the stick on top of the withers, as shown here.*

*This donkey has a malocclusion of the teeth known as an underbite. Because this trait is hereditary, you should not buy an animal such as this one for breeding stock.*

noses, untended sores, overgrown feet, and coughing. All of these can be indications of poor husbandry and potentially chronic health issues.

Ask the seller about his or her vaccination and worming program. Ideally, the seller will have written records of both to send home with you. Check the donkeys' bites (teeth alignment). Malocclusion (faulty alignment) is an inherited trait and a critical issue with breeding stock. If serious enough to make eating difficult, it can cause problems for a pet or performance donkey, as well.

## Check Out Breeding Stock

If you're buying breeding stock, you have to consider several aspects. For instance, when choosing a jack, make sure he has two testicles. Sometimes it can take up to two years for jacks to "drop" their testicles, but don't buy one older than

that unless his "manly parts" are clearly down. For bred jennets, ask the seller if he or she has checked them in foal. If so, was it done by ultrasound or palpation? Ultrasound is preferable because it will reveal whether a jenny carries twins. Palpation will not do so.

Twinning, which is extremely common in Mammoths, is dangerous. Many times carrying twins will cause a jennet to abort late in pregnancy; if she does carry to term, the delivery can be difficult, often resulting in the loss of the foals and sometimes even the jennet herself. Ultrasound reveals twins early enough that reduction can be performed if necessary. (Reduction involves pinching off one or both embryos—preferably one, but both if the embryos are overlapping or too close together.) Ask if the seller offers a guarantee of pregnancy and, if not, whether a free rebreeding is offered.

# Getting the New Donkeys

Our experts share advice on purchasing new donkeys
and methods for transporting them home.

### Incredibly Shrinking Donkeys

"My suggestion on buying is buy from a reputable person or have a vet check the animal before [you make] the trip if you are going any distance. [The vet] may charge $100 for a trip call and an inspection, but that will not buy much gas and can save you a lot of time. I once drove over 600 miles *one way* to look at two Mammoth jennets with great pedigrees. They each shrank about four inches in the day I was on the road!"

—Joe Thomas

### Homework and Heart

"Do your homework on buying your first one, find out what type you like/want, and get the very best you can afford. Sometimes all the practical things/reasons don't line up with what your heart says. *Sometimes*, you have to follow your heart."

—Bobbi Ward

### Seeing It All

"A buyer should be able to see the animal being caught, haltered, tied up, brushed out, teeth checked, feet picked up all the way around, a saddle or harness set on his or her back to see if there is any reaction, even being led around. Even ridden or driven if trained to do so. Or told that nothing has ever been done and you'll be starting from scratch before you make the trip. [And] I like to see conformation photos before I go."

—Kristi Kingma

### Green-on-Green Bruises

"Green on green equals black and blue. Not that it wouldn't be possible for a green donkey to work out for a green rider, but it generally works out much better if at least one of the pair has some experience"

—Kris Anderson

### Loading-Time Allowance

"When picking up the animal in a trailer, allow an extra hour to load, because surer than the cat had kittens, that animal has never seen a trailer, let alone ever been in one."

—Jim Ensten

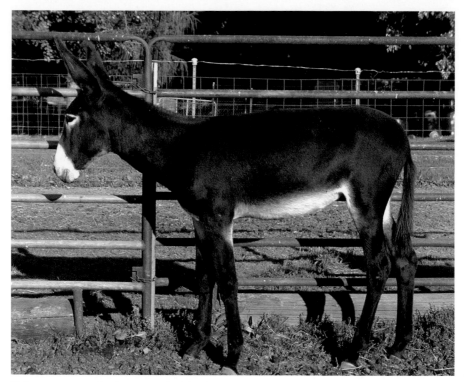

*This young jenny displays a roached back and a ewe-neck, both of which make her conformation undesirable as a breeding prospect.*

Tell the seller you would like to see offspring and other relatives of the donkeys you're looking at to get an idea of their temperament and conformation. Note the attitude of the ones you're thinking of buying. Most donkeys are friendly and sociable. Even the ones that are a little aloof generally come around in time. But if your prospects take off, snorting and wild eyed, for the back forty, you should think twice about buying.

## See Trained Donkeys Perform

If you're buying a trained donkey, ask to see the animal perform under saddle or to a cart. If possible, try riding or driving the donkey yourself. Donkeys tend to take a little time getting to know a new handler and may not behave flawlessly for you under these circumstances, but you should be able to get a feel for an animal's level of training and performance.

## Getting the Paperwork

Once a deal has been struck, ask for any guarantees the seller offers in writing. If you have problems with the donkeys later, a written contract can go a long way toward resolving any issues. Request any written health records the seller has or at least the dates of the most recent vaccinations and deworming. Find out if a current, negative Coggins test is available. In virtually all states, an owner must have one to sell an equine. In most places, you must have a health certificate (good for thirty days) to transport an animal across state lines. Discuss who pays for these items. Generally, the seller provides them, but a few sellers will try to pass the costs along to the buyer.

If the donkeys are registered, ask for a signed transfer form and the registration certificate, along with a breeder's certificate if you're buying a bred jenny. These will enable you to transfer the animals to your name. Occasionally, a seller (especially with higher-dollar sales) will take care of this and send all necessary transfer paperwork to the appropriate registry for you. Registrations and transfers can sometimes take eight to twelve weeks or even longer to process. A quick note or call to the organization's office to check on things generally finds the paperwork completed and on its way back to you.

## Transporting and Settling the Newcomers

If you do not have your own truck and trailer, ask the seller for advice. Sometimes he or she might be able to haul the donkeys to your home for you or give you names of reliable transport companies.

When your new donkeys arrive, give them time to settle in. If possible, quarantine them away from other equines (at the very least limiting nose-to-nose contact) for ten to fourteen days. Monitor appetite and behavior. Sometimes donkeys are reluctant to drink water in a new place; adding a little sweet flavoring, such as Kool-Aid, can help in the adjustment. If you can acquire some of the grain and hay they have previously been fed, blend it with your own over the course of a few days to avoid digestive upsets. Use treats, scratching, and lots of brushing to make friends with your new longears. This generally is not difficult to do. The hard part will be escaping from your new friends, who will try to convince you to quit your job and continue with all the attention 24/7.

Check with your veterinarian to find out what kinds of vaccinations he or she recommends for your area. Compare this with the health records supplied, and have any needed shots administered. It's also a good idea to deworm the donkeys if it has been six weeks or more since the previous administration.

# EIA (Coggins) Test

Equine infectious anemia (EIA) is an incurable disease transmitted between equines by bloodsucking insects. A test for EIA (commonly referred to as a Coggins, after the veterinary virologist, Leroy Coggins, who developed it) is usually performed yearly. Testing requirements, however, vary by state, so check your local laws. In most places, you must have an EIA test done before you take your donkeys to shows, on trail rides, to clinics, or to any kind of gathering of horses, mules, and donkeys. You must also have one done before selling a donkey. You have to keep an up-to-date negative EIA test certificate with you when traveling with your donkey. In addition to the negative Coggins certificate, most states require a health certificate (good for thirty days) to cross their borders. This is needed regardless of whether you plan to stop or unload in that state. Authorities levy stiff fines and penalties for failure to comply, so have your paperwork in order!

# chapter FOUR

# Housing and Feeding Your Donkeys

**H**ousing and feeding donkeys can be an incredibly complex subject. There are many differing opinions and dozens of ways to accomplish the tasks. But ultimately, it can be stripped down to some fairly basic principles. Donkeys need shelter from wet and windy weather (particularly when combined with cold temperatures). They also need protection from biting insects. They need plenty of fresh, clean water, kept thawed during freezing temperatures. Adapted from the semiarid regions of North Africa, donkeys thrive on diets higher in fiber but lower in quality than those eaten by horses. Donkeys also need *much less food* than horses do. Probably the biggest mistake donkey owners in North America make is overfeeding.

## Home Sweet Home

If you're bringing donkeys home and live on a piece of rural property, chances are you already have some sort of structure that can be easily modified for the new arrivals. Donkeys are not picky about their "digs"; they just want to be able to get in out of the rain, snow, and wind. They do prefer to keep their options open and are not crazy about being kept stalled. In their perfect world, they can mosey between shelter and pasture at will.

If you don't have a structure available, you can construct a fairly simple one. The easiest and most economically feasible building is a three-sided shed made of whatever materials are readily available and least expensive. Before you start construction, check with your local zoning officials to find out whether you need any permits. Fortunately, a portable three-sided shed, put on skids so you can drag it to different locations, is not usually subject to building codes. This moveable structure can also offer some future flexibility in your housing plans.

*A simple, three-sided shed such as this one provides adequate shelter for donkeys. Built on skids, the shed can be easily moved to a new location.*

If you do choose to construct a donkey shed, keep the following points in mind:

- Locate shed on high ground, with the back side against prevailing winds.
- Make sure there is good drainage all around the building, with room for gates where needed.
- Make the ceilings high enough that the donkeys won't bump their heads.
- Think of access and proximity of water, hay storage, and electricity to shed.
- Have a suitable area nearby for manure removal.

Other important housing aspects to keep in mind are the need to provide the right type of floor and proper bedding and the necessity of using chewproof materials.

## Comfy Floors and Bedding

If at all possible, keep the floors of sheds or barns earth. Concrete can be hard on the feet and legs. Packed clay and dirt are best, although these will occasionally need to be refilled as they will become dug out with repeated cleanings. If you want to keep this from happening, put down thick rubber stall mats. Initially an expensive outlay, the mats pay for themselves with the added ease of mucking out sheds and stalls.

If the donkeys are to be confined for any length of time, they will need some sort of bedding, as well. Straw, wood shavings, sawdust (excluding black walnut, which is toxic), ground cobs, and similar materials all make acceptable bedding. Use stall "sweeteners" on urine spots inside the shed or stall. This will keep odors down and help control ammonia fumes.

## Beaverproof Materials

When it comes to wood, either for buildings or fences, donkeys are very much in touch with their inner beaver. They will eat the wood almost as fast as you can put it up, although they are not quite as destructive on rough-cut oak lumber as they are on soft woods such as pine. For this reason, attach metal flashing to all unprotected wood edges and corners and install an electric hot wire 4 to 6 inches inside the perimeter of wood fencing.

## Out to Pasture

As with the majority of equines, the most natural and healthiest place for a donkey to be a donkey is out on a pasture of some kind. It doesn't have to be a big area (unless you expect the pasture to fulfill all of your donkeys' nutritional needs). An area anywhere from the size of a round pen, 50 to 60 feet in diameter, to several hundred acres will give the donkeys a chance to exercise and enjoy fresh air and sunshine.

Depending on the size of the donkey and on its nutritional demands, pasture may be all that is necessary to sustain the animal during the growing season in North America. In most regions, for Mammoths, it takes one to two acres per donkey to provide grazing with no supplemental feed. Standards and Miniatures require less.

However, this pasture should not be one to two acres of weeds and dirt. Donkeys will starve on small lots with nothing but noxious weeds that have been defecated and urinated on. They do not require the lush, well-fertilized pastures horses prefer, but theirs should be healthy, free of bare patches (except the donkey-constructed paths and dust holes), mowed to keep weeds down, and harrowed occasionally to break up and scatter manure. If you expect the pasture alone to provide all required roughage, your donkeys need to graze about six hours a day to consume adequate amounts.

*Ishtar, a standard donkey, grazes contentedly in a leaf-strewn, brushy pasture. Lush grassy pastures are not favored grazing areas for donkeys.*

# Room and Board

Here our experts offer some advice on how to
properly house and feed your donkeys.

### Keeping His Head Down

"If you are going to include
some advice on containing
a jack, I would suggest that
the fence should be high
enough that he cannot get
his head over it. If he can
get his head over the fence
and get his chest against it,
he can/will push it down or
jump it. I would also suggest that he have an out-
side paddock of at least an
acre or so. I've never had a jack with a
'jack sore,' which I attribute to plenty
of fresh air and sunshine and a clean,
well-kept stall."

—Karl Schneider

### Preventing Toothpick Barns

"One piece of advice we should have
taken to heart immediately after getting
our donkey was to protect all chew-
able surfaces! Our once-pristine barn
has been chewed like an old toothpick!"

—Terry Lupien

### Remembering the Differences

"Things vary in different locations.
It's up to the donkey owner to deter-
mine the prevailing winter winds and
which way the shelter should face,
which variety of hay [to use], [which]
toxic plants [and] parasites [to watch
for], and what vaccinations should be
given. There can be big differences
over fairly small distances."

—Michelle Moss

### Listening to Northerly Advice

"One thing that I would do differently
with the barn is to make sure the stall
doors aren't on the side of the barn
where the snow comes off. Once that
snow comes down off the roof, it is
very hard to move; even with a snow
blower mounted on a tractor it seems
to set up like concrete. On a metal
roof like we have, you can put snow
stoppers on the roof about 3 feet back
from the edge of the roofline, and they
do hold the snow back. They work so
well that the snow on the north side
of the barn will come off before the
south side, which gets the sunshine
(when it shines, which isn't often in
the winter). Inside the stall, all wood

edges should get a metal edge; corners for drywall work well. Salt blocks can be left in the stall, and they will last longer than when left outside for the weather to wear them down."

—Larry Thompson

## Using Hog Fuel

"To help with the mud here in Oregon I use 'hog fuel,' which is processed cedar bark. It works very well. I use it around the barn or any high traffic areas, gates, shelters, etc."

—David Anthony

## Making the Drainage Grade

"We have gravel in our pens, and the only time it gets soft is in the spring when the frost comes out of the ground. The pens are sloped though, and if we get low spots, we grade them so they'll drain. It's wet here in Massachusetts, and we live at the bottom of a valley, so we have to grade the pens and around the barns or they'd be mucky messes. One thing we have figured out is that if we let the hay and manure build up on top of the gravel, then it'll get sloppy again. We try to avoid scraping right down to the gravel so we don't scrape too much of it back up again, but we do scrape it often enough to prevent it from getting sloppy. We built one run-in at a bad location (too flat) and then had to build up inside and around the front of the shed so the water would drain away. It doesn't take a particularly steep slope to get water to drain, but it does need to be graded off, and then the hay and manure need to be scraped off it."

—Kris Anderson

With their varied palates, donkeys have much in common with browsing species such as deer and goats. Donkeys like to alternate grazing on grass with nibbling blackberry brambles, saplings, dead tree bark, and roughage that would seldom interest horses. You'll find donkeys particularly well suited for wooded pastures with brush and undergrowth, as this is a culinary luxury to them.

To keep your pastures in a donkey-approved state of production, try to design several smaller paddocks or lots for rotational grazing. This, along with harrowing the manure, will keep pastures healthier and stem the creation of weedy, overgrazed areas. Be prepared to give up several spots in your pastures to "donkey holes." No donkey habitat is complete without its dust craters. Donkeys carefully construct these craters for luxurious napping and dust applications. Particularly during the summer months, they will line up for a turn at rolling in the dust hole.

## Fenced In

Fencing, like housing and feeding, can be as simple or as complicated as you choose. Unfortunately, the better looking the fence, the higher the price. If you consider aesthetics important, you can contain your donkeys behind post

## DID YOU **KNOW**

Donkeys can withstand up to 25 percent weight loss from dehydration and recover the loss when water is available. A donkey's resting metabolism is 20 percent lower than that of a horse; energy requirements may be 25 percent lower.

*Lulu, a Mammoth, models a grazing muzzle. Although a donkey can still eat and drink while wearing a muzzle, it slows consumption and helps prevent obesity and its complications.*

# Overfeeding

Three of the most serious illnesses afflicting donkeys can be the result of overfeeding.

**Epiphysitis**—this is a condition associated with rapid growth. Their large size makes Mammoths particularly susceptible. Swellings inside the fetlocks and knee joints, often causing lameness, should make the owner suspect epiphysitis. Sometimes, reducing the food intake and/or content will help. If not, anti–inflammatory medication may be required.

**Hyperlipemia**—this usually occurs after the donkey has gone off its feed from some stressor and is common in obese and, especially, pregnant animals. It causes an excess of fat in the blood and is life threatening.

**Laminitis**—also known as founder, this is an often-chronic inflammation of the sensitive layers of the hooves. The donkey's feet will be warm to the touch, and the animal will be lame, often standing rocked back on its heels.

*These Miniature donkeys stand behind a fence made of cattle panels. Sturdy, flexible, and easy to move, cattle panels are great if you need to reconfigure lots.*

and board, pipe, or vinyl/PVC fencing. A utilitarian permanent, if a less visually impressive, choice is the wire field fence, usually some 48 inches high, with square openings in the mesh. When the fence is stretched tight, as it's intended to be, the mesh makes a formidable barrier against dogs or predators and may be a particularly good choice for fencing Miniatures as their small size makes them especially vulnerable to attack.

Cattle and/or hog panels are semirigid, 16-foot panels with square and rectangular openings. Cattle panels are 50 inches high, and hog panels (more suited for Miniatures) are 34 inches high. Their main benefits are that they don't have to be stretched, don't sag like wire fencing, and are portable.

The less expensive, least attractive, but by far most versatile fences are the electric ones. They come in tape, rope, and plain smooth wire. Financial savings aside, electric fencing offers the ability to move and refashion grazing areas as needed. This can be a great boon in rotational grazing. Donkeys are extremely respectful of electric fencing and can often be contained with as little as one strand. Even if electric fencing is not used as the primary encloser, it is a terrific backup to help protect other fencing (and avoid the "beaver" syndrome that your expensive board fences may fall prey to). It will also keep donkeys from mashing field fencing by rubbing itching behinds on it and keep jacks from thinking about jailbreaks.

Barbed and high-tensile wire are generally *not* the best choices for equines,

# Toxic Plants

Most equines will avoid toxic plants or at least not ingest them in quantities that cause illness. However, sometimes, due to excessive hunger, curiosity, or accident, they may sample one of these plants. Some of the more common toxic plants are listed below. Not all plants listed here are deadly, but if in doubt, call your vet.

| | | |
|---|---|---|
| Alsike clover | Horse chestnut | Sheep laurel |
| Apricot tree[1] | Horseradish | Peach tree[1] |
| Azalea | Horsetail | Plum tree[1] |
| Baneberry | Hyacinth | Pond scum (algae) |
| Black locust | Iris[3] | Potato[6] |
| Black walnut[2] | Jack-in-the-pulpit | Privet |
| Box wood | Japanese maple | Ragwort |
| Bracken fern | Japanese yew | Red maple |
| Buckeyes | Jimsonweed | Rhododendron |
| Buttercup | Larkspur | Rhubarb[7] |
| Cherry tree[1] | Leafy spurge | Rubber tree |
| Chives | Lily-of-the valley | St. John's wort |
| Christmas rose | Lily's bulb | Star-of-Bethlehem |
| Cocklebur | Lupine | Sensitive Fern |
| Corn cockle | May apple | Skunk cabbage |
| Crimson clover | Monkshood | Sudan grass[8] |
| Daphne | Mountain laurel | Sweet pea |
| Dutchman breeches | Mushroom | Thorn apple |
| Elderberry | Narcissus | Tomato[9] |
| Ficus | Nightshade | Vetch |
| Foxglove | Oak[4] | Water hemlock |
| Foxtail | Oleander | White snakeroot |
| Ground ivy | Onion[5] | Yew |

1 Twigs and foliage can be lethal; antidote is effective if given very quickly.
2 For bedding, be careful to get shavings that contain no black walnut.
3 The fleshy root is lethal.
4 Acorns in large amounts cause gradual kidney damage in livestock.
5 A few are fine, but large amounts for several days will destroy red blood cells.
6 Leaves and vines can be lethal; the normal, healthy tuber itself is harmless.
7 Leaf can be lethal.
8 Under 18 inches, dark green; taller grass is less toxic.
9 Vines may be poisonous.

*Keeping grain in a rodent-proof metal can is an ideal way to store it. Grain left in bags will soon be riddled with holes or torn apart by raccoons, possums, rats, and mice.*

although donkeys don't have as many problems with it as horses do. Because of their nervous, high-strung dispositions, horses can spook and run into barbed or high-tensile wire, sustaining serious injury. This is fairly uncommon in the laid-back donkey but still a possibility.

When contemplating fencing choices, assess needs. The different sexes obviously won't require the same features. Jacks need extremely strong, durable fencing, preferably lined with at least one strand of electric wire. Foals need protection from predators and fencing that keeps them from rolling out underneath it. For larger donkeys, most fencing should be 48 inches or higher. Keep wire fencing stretched tightly to avoid injuries from pawing or getting feet tangled. Fences should be smooth on the inside, with no sharp protruding wires. Fasten boards and wire to the inside (animal side) of posts.

## What's on the Menu?

As stated earlier in this chapter, the biggest issue associated with feeding donkeys is fighting the tendency to overfeed them. The donkeys themselves will encourage overfeeding, braying mournfully and batting sad, hungry eyes as their miserly owners dole out a handful of oats. Donkeys will have you believe that they just barely found the strength to stumble up to their feed pans and are in danger of starvation. Don't buy it.

The obese or soon-to-be-obese donkey gives itself away by sporting a very thick, hard roll of fat along its neck and "pones" of fat on its back. The neck roll, if left unchecked, will eventually fall over, causing a permanent, unsightly disfigurement of the animal referred to as a "fallen crest." For the majority of Miniature and standard donkeys, life is one long Weight Watchers experience. Mammoths, however, actually tend to go through a gawky, gangly growth spurt between one and three years old, when they resemble a teenage boy with a hollow leg. Keeping them nutritionally fulfilled without going too far and causing joint problems can be difficult. Thus, feeding donkeys is a very individualized process, dependent upon a donkey's breed, age, intended use (such as maintenance, breeding, or riding), and specific metabolism.

## A Gut for Grazing

Like all equines, donkeys are grazers and do best with many small meals throughout the day. The donkey's gut operates most efficiently by having a steady intake of small amount of foodstuffs to process.

Any changes to the diet need to be made gradually. Feeding at the same time each day helps avoid digestive upsets.

## Hot and Sweet Feed

Donkeys do not need a whole lot of protein, but growing animals (Mammoths especially), pregnant/lactating jennies, and working donkeys do require adequate amounts. In determining how much to feed, you need to measure the food by weight rather than volume. Weight of grains varies considerably. For instance, a quart of corn weighs nearly twice what a quart of oats does. In general, donkeys don't need much grain, but in cases such as those listed above (growing, pregnant, working), feed them oats. Oats have a good amount of protein but low digestible energy. In addition, they are very fibrous and add bulk to the diet.

Corn, a common grain choice, has high digestible energy but low fiber. It must be fed more carefully as it is a "hot" feed—that is, one that tends to cause obesity if used in excess. Either plain oats or a 10 percent "sweet feed" (a combination of oats, corn, barley, soybean meal, salt, and minerals, bound together

| General Guidelines for Daily Feeding | | |
|---|---|---|
| Breed | Maintenance | Working* |
| Miniature | 0-1 cup grain<br>4-5 lbs. hay | 1-2 cups grain<br>5-6 lbs. hay |
| Standard | 0-1 lb. grain<br>7-8 lbs. hay | 1-2 lbs. grain<br>9-10 lbs. hay |
| Mammoth | 0-2 lbs. grain<br>14-16 lbs. hay | 4-5 lbs. grain<br>16-18 lbs. hay |

*Working constitutes growing, pregnant, lactating, being ridden, driven, and other such activities.

*Mammoth donkeys Perty and Tookay belly up to the hay on a cold winter's day. Although feeding off the ground is safe when it's frozen, it's not a good idea in general.*

with molasses), manufactured by a feed company or a local grain elevator, is sufficient for most donkeys and a much better choice than straight corn.

## Healthy Hay

Ninety percent or more of the donkey's diet should be composed of forage, either hay or pasture. Hays are classified as legumes—alfalfa, clover, lespedeza—or grasses—Bermuda, brome, coastal, fescue, orchard, timothy. These represent the primary North American varieties. There are regional preferences, and other types of hay than those listed may grow best in certain parts of the country. Legumes, in general, are richer

and more fattening than most donkeys need. Donkeys do best on one of the grasses or a grass mixed with a bit of clover or alfalfa.

The primary characteristics to look for when you purchase hay are good color (generally, the greener the better), cleanliness (as little dust as possible and no mold or foreign objects), and fine stems. The fine-stemmed issue is not as important for donkeys as it is for horses. First-cutting hay tends to be a little weedier and coarser stemmed than later cuttings, and horses may find the former to be unpalatable. Donkeys are not as picky in that respect. Farmers usually put up three to five cuttings of alfalfa, or alfalfa-mix hay,

*Hay should be green and sweet smelling, not moldy and musty like this. Break open a sample bale and check the quality before purchasing large quantities.*

*This good, green, fine-stemmed Bermuda grass hay is an ideal choice for easy-keeping smaller donkeys. Legume hays are best fed only to Mammoths.*

and one or two cuttings of grass. They will either bale in small squares (50 to 60 pounds), big squares (600 to 800 pounds), or round bales (600 to 1,200 pounds). It's up to each owner to choose which type he or she can best handle, store, and feed, but don't use the big round bales unless you have a group of donkeys that will clean the hay up within or a week or so. Donkeys will usually pick through the outer weathered, spoiled layers of round bales to get to the good, dry hay underneath. If only one or two donkeys have an entire bale to themselves, it will decompose faster than the animals can eat it.

Be leery of buying any hay that has been rained on after cutting but before baling. Even if it dries well before it's baled, it usually loses most of its color (and therefore its vitamin A). Its leaves become brittle and fall off, leaving only stems. The food value is virtually nil.

Take a handful of hay to test. If you bend it and it snaps, look elsewhere.

You should buy as much of your hay as possible from the same source. This will keep the donkeys' diet unchanged and consistent, which always helps avoid digestive trouble. A good hay provider, like a good veterinarian and a good farrier, is a valuable resource.

## The Right Mineral Block

Donkeys' mineral needs can be met, for the most part, with good hay and grain and a trace mineral salt block. Be sure the block is *salt*, and not one of the soft, crumbly mineral/protein blocks. Donkeys will inhale those like candy. A salt mineral block is usually dark red in color and very hard and weighs about fifty pounds. Don't use the protein tub "licks" manufactured for horses, which are too rich and fattening for donkeys.

Check with your local extension agent about possible selenium deficiencies in your region. Selenium is a mineral that can cause serious problems if it is over- or underfed to a donkey, so seek expert advice on using it.

## Feeding Tips

Here are a feeding tips to help keep your donkeys healthy and hardy:

- When turning donkeys out on early springtime pasture, let them fill up on hay first and limit grazing to an hour or two. Build up slowly.

- Make sure feeders are clean and have no hazardous edges. Don't let feed build up in feeders and get spoiled or moldy.
- In general, do not feed donkeys off the ground. Ingesting sand and dirt can lead to colic and to heavy parasite infestation.
- Expect each donkey to drink eight to ten gallons of water daily. Keep water thawed in winter, and if using stock tank heaters, protect cords from chewing by keeping them out of reach or enclosing them in chew-proof material, such as PVC pipe.

The owners of these Miniatures at Hester Farm in Missouri put their salt block in a tip-proof feeder. This mineral is crucial for donkey health.

# chapter five

# Donkey Behavior and Handling

The donkey, in the depths of its soul, is a cautious, conservative, careful individual. It will thoroughly examine the safety quotient of each endeavor. The donkey will not be hurried, nor will it let anyone make up its mind for it. Because of this, the donkey has earned an unfair reputation for being stubborn, bull-headed, stupid, and obstinate. Rather than being respected for its deliberation in an uncertain situation, the longear is frequently ridiculed. Those who get to know donkeys will quickly come to appreciate their innate intelligence and understand why donkeys may take a little longer to decide but always make the right choice in the end.

Unlike horses, which have a well-developed flight instinct when startled, the donkey's instinct is to freeze. The donkey will usually trot or run only a short distance before turning to assess the "danger." Donkeys are not being willfully obdurate when they freeze as you try to lead them across running water. Thousands of years of ancestral knowledge from the deserts where they originated are warning these animals to tread carefully through unnatural terrain.

Being instant-gratification junkies, humans tend to have scant patience with the donkey's cautious and thoughtful nature; we blame that nature for not providing us with a speedy and unquestioning response. To get along with our longeared friends, we need to slow our pace. We need to be on "donkey time."

## Donkey Dislikes and Likes

Learning about the donkey's unique personality traits can provide us with a better understanding of what makes this animal tick and help us in our handling. In this chapter, we will discuss a number of the donkey's unique habits and behaviors as well as its dislikes and its likes.

*Mammoth jenny Mae checks out Purrcy the cat. Donkeys can learn to tolerate an owner's pets but may chase strange animals away.*

## Dislikes

Donkeys have an inherent dislike of small animals. This trait makes them useful as guard animals, protecting species such as goats, sheep, and cows from dogs, coyotes, and other predators. Care must be taken, however, lest the donkeys decide to stomp their own charges. Introduce a donkey and its charges slowly and cautiously. Give smaller animals an escape zone. Keep an eye on pets. Cats and dogs (not to mention poultry and other small animals) are not exempt from the donkey's territorialism.

Most of the time, donkeys simply put their ears back and enjoy a good game of chasing their small prey, but sometimes, in the case of an aggressive jack or a maternal-minded jenny, a donkey means business. Keep an especially close eye on infants and toddlers. Donkeys are usually absolutely wonderful with children, but as in the previous scenario, there are certain situations that merit extra caution.

To a donkey, cracks in the pavement, odd shadows, painted lines down the center of roads, enclosed horse trailers, and bodies of water are all highly suspicious. In their minds, they have perfectly good reasons to be dubious about walking over, into, or through all of those things. It is your job to negotiate and convince them to do so.

## Likes

Donkeys are social, gregarious creatures, always happiest with fellow donkeys. They travel throughout the pasture in single file, "conga" lines. Although donkeys will not willingly take a water bath, they love to bathe daily in their dust craters. At times, all a person may see is a cloud of dust obscuring the contented donkey beneath, rhythmically beating his tail to keep the dust wafting over him. Unlike other grazing animals, a herd of donkeys will not make shambles of a pasture. Instead, they tend to be fairly neat, urinating and defecating in certain areas. They will tread well-worn paths rather than leave an entire field pockmarked by hoof indentations.

## Taking Donkeys in Hand

Whether they are being used as pets, for breeding, or as working animals, all donkeys need some basic handling and training. Donkeys must be able to be handled by you, the owner, as well as by a farrier for hoof maintenance and a veterinarian for routine and emergency health care. At the *minimum*, you should train every donkey to be touched all over, including having it feet handled, to stand tied quietly, to lead, and to load in a trailer.

Training methods vary enormously. The following three training methods are the ones most commonly used.

*Bribing*: This is training with some sort of treat as a reward, usually food, though sometimes just scratching or verbal praise. As with any sort of animal training, the key element is timing. The donkey must associate the treat with the correct behavior.

*Pressure/release*: Just like it sounds, in this training method, you put pressure on the animal in some form (such as pulling forward on a lead rope to get the donkey to take a step toward you or pushing him in the flank with a finger to get him to move to the side). The instant you feel or see "give" from the donkey, you release the pressure. The release is his reward, and the timing of it is of the utmost importance. Just an added word about this particular method and donkeys—donkeys tend to be resistant to pressure. They push into it rather than moving away. They are actually easier to drive forward than to lead.

*Pressure from behind*: When you lead donkeys, they will sometimes "lock up." By nature, donkeys resist being pulled forward. If you don't do something to get the longear moving, a tug-of-war soon ensues. If your donkey locks up, step back to the side and slightly behind while clucking or tapping the donkey lightly with a whip; this can get the stalled-out animal

*Gus takes a luxuriant roll after being unsaddled. Donkeys love to work dust and dirt into their coats.*

*Noah and Rachel prepare for a leading lesson. Noah will tap her hocks with the whip while giving her a verbal command to move forward.*

# Don't Leave Halters on

It is a very dangerous practice to leave a halter on an unattended donkey. Many times, folks are tempted to leave the halter on an animal that may be a bit skittish or hard to catch, but there are a lot of hazards associated with this practice. Donkeys are forever rubbing and scratching their heads against inanimate objects. All it takes is hooking part of the halter on some protrusion, and the wreck is on. Generally, donkeys aren't nearly as extreme in their reactions to this kind of situation as a horse, but young donkeys, in particular, can struggle until they've suffered significant injury, including broken necks. Donkeys are also flexible enough to reach forward with a hind hoof to scratch their heads. Occasionally, that back leg will become entangled in the halter with similarly disastrous results. It's best to only have a halter on your donkey when you are present and using it to lead or tie.

to achieve forward momentum. Be sure to keep an eye on the "kicking zone."

All three of these methods or some combination thereof have proven successful in training donkeys. Sometimes a bit of experimenting is required, and what works for one donkey may not be as effective on another. Whatever methods you employ, the most important aspect to keep in mind is safety (not getting stepped on, run over, kicked) for trainer and trainee. Keep lessons fairly short, and always quit on a good note. Donkeys are actually very quick learners and in most cases are happy to perform what's asked of them.

## Touching

In general, donkeys have no problem learning to accept human touch. They quickly recognize the benefits of forming a tight bond with the provider of the food. When you're dealing with a leery or standoffish donkey, one of the fastest ways to make friends is with food. Feed the donkey in a small enclosure, and speak reassuringly to it. Work your way closer until you can give him a pat or rub, then back away. Repeat this until the donkey realizes there is nothing negative about this situation.

Too often we only catch up or halter our donkeys for some activity they're not especially fond of (such as a vet visit, a farrier visit, or body clipping). Your donkeys need some petting and scratching with no ulterior motives attached; if they don't get this, they may start deciding to "get out of Dodge" when they see you coming with a halter and a lead rope.

With an extremely flighty or fearful donkey (such as an adopted Bureau of Land Management burro), some folks have had success with the "broomstick method"

*Eeyore accepts a little lovin' from Noah. Mostly, donkeys and children go together like peanut butter and jelly; just make sure no aggressive jacks or maternal-minded jennies are nearby.*

# Hair Today, Shorn Tomorrow

Body clipping your donkeys can be desirable for many reasons. Left to Mother Nature, a donkey retains most of its long, dead coat well into summer. When moisture from rain and sweat gets trapped between layers of hair and the skin, a donkey may get a fungal infection or become infested with lice. Donkeys that participate in performance events may sweat heavily, and of course, any donkey going to a public event looks more "spit polished" with a fresh haircut.

*Shelby displays the typical long, heavy coat of a weanling or yearling donkey. Without clipping, remnants of this coat may not shed out until she is two years old.*

For the larger donkeys, most people use large clippers to do the majority of the body and a smaller set for the head and ears. For Miniatures or standards, the smaller clippers may be enough for the entire job. Before you start, clean and dry your donkey. A dirty donkey will dull blades at an alarming rate.

To easily restrain your donkey while you work, cross tie it between two solid posts, with the lead ropes snapped to the side rings of the halter. Before you approach the donkey to start cutting, turn the clippers on and let it get used to the noise.

Clip against the hair in smooth motions, using your hand to gently stretch the skin taut in delicate areas such as the chest, armpits, and flanks. Keep the clipper blade as flat against the body as possible, and adjust the direction whenever the lie of the hair changes. Make sure you stop fairly frequently to dip the blades in a coolant specially made to clean and lubricate them.

Unless the donkey is going to be shown, leave the hair on the legs. Flies love to chew on donkey legs,

*Fresh from a complete body clip, Shelby is looking good! Her skin will be healthier, as well, now that sunlight and air have better access to it.*

making life miserable for the animal in early summer. The added protection of the hair can afford some relief. If you live in a climate where weather is still cool when you cut the hair, you may need to blanket the donkey for a few weeks.

of touch. You use the broomstick as an extension of your arm. Gently, slowly, rub the animal all over with broomstick to desensitize it to touch. From there you progress to touching with your hands.

## Tying

Unlike horses, which sometimes rear or pull back and break halters, donkeys usually don't have too many issues with standing tied. If there's going to be any tying fireworks, they will probably occur when you train a foal or a feral burro. The first time or two they are tied, babies and wild donkeys will leap about, doing gymnastics.

For this reason, when training one of these animals, always pick a safe location with nothing nearby that could injure a struggling animal. Always tie the animal to a secure post and not to a cross board. Tie the donkey fairly short (no more than 12 to 18 inches of lead rope between animal and post). Tie at wither level or higher. If possible, use a stretchy type of training lead rope or a regular lead rope tied to an piece of tire inner tube, which you affix to the post. The give will help lessen the donkey's shock at being restrained. Use a quick release type of knot, and always keep something handy to cut the rope in case of an emergency.

## Leading

Teaching leading can prove challenging, mostly, as noted above, because of the donkey's tendency to brace against pulling pressure. To make the training easier, use two people, one placed fore and one aft. The animal will move away from the person pressuring from behind and quickly learn what's expected. When you train alone, line the donkey up between you and something solid, such as a fence or the side of a barn, to keep the animal next to you as you work. Stand at the donkey's left shoulder, holding the lead rope in your right hand. With your

Noah and Penelope practice their leading exercises. Many Miniature donkeys are small enough and docile enough for even children to train.

left hand, hold a long whip to be discreetly flipped around behind your back to tap the donkey's rump or hocks. This should promote forward motion at the same time you cluck, smooch, or give a verbal command such as "Walk on" or "Let's go."

*Be forewarned*: When they are worried about something, donkeys have a habit of taking their handlers for a drag. Although, they're not being malicious, they can prove unstoppable when they brace their muscular necks and take off. Even when a donkey's only moving at a walk, few people can halt or change the direction of a determined donkey. To keep control, use a stud chain when teaching leading. This is a length of chain usually 12 to 18 inches long, resembling a chain dog collar. You snap this to your lead rope, then run the chain through the rings of the halter, under the chin, and fasten it to the ring on the "off" (right) side. It gives an added measure of control and can be very

*Brigitte is none too happy to have the farrier trimming her hooves. No matter how your donkey feels about it, this trimming needs to be done at least two or three times a year.*

helpful when you've got the donkey in a new place or situation in which you feel the animal may be especially nervous.

## Feet Handling

Donkeys tend to be fairly conservative when it comes to folks fondling their footsies, but of course, it's a necessity that you be able to do so. Long, overgrown feet are one of the most frequently observed forms of donkey neglect. Because donkeys sometimes put up quite a fuss over foot handling, many owners simply let the feet go uncared for. This can lead to chronic, painful, long-term lameness. Ideally, the breeder will start picking up the donkey's feet in foalhood, rubbing, tapping on them, and handling them thoroughly. When treated in this fashion as foals, donkeys seldom grow up giving their humans any problems.

On a regular basis, the feet should be cleaned out using a metal pick designed for this purpose. Tie the donkey "short and high" in a safe working zone. Stand next to the donkey's shoulder, facing the rear. Run your left hand down the front leg, at the same time using the words *lift* or *give* as you pick up the hoof. Clean thoroughly with the pick, and then set the hoof down, praising good behavior.

Move on to stand next to the back haunches, still facing rearward. Again, you should speak to the donkey, patting it gently on the hindquarters and making sure not to startle it. Slide your hand all the way down the hind leg, and repeat the command for the donkey to give up control of the foot.

As you're handling the rear feet, always be aware of the animal's movements. Stay parallel with its side so that, if necessary, you can push directly away and out of kicking range. Watch and feel for signs of muscle tensing, which can be an early warning that the donkey is getting ready to misbehave. Always keep in mind where your vulnerable body parts are in relation to donkey feet. Donkeys are not kickers by nature; they're more likely to just try to jerk their feet out of your hands. If pressed too hard, however, a sufficiently scared donkey will sometimes kick.

Work your way around the donkey until you have cleaned all four hooves with the pick. If you have trouble getting the donkey to go along with any of this, stop what you're doing and work on desensitizing the animal. A soft cotton rope can be gently draped around a pastern (ankle) and used to lift the foot. A cane or stick with a hooked end can be used the same way. When the donkey becomes relaxed and familiar with having its feet lifted in this manner, you can progress to using your hands to complete the task.

At least a couple of times a year, sometimes more frequently depending on the rate of growth, the donkey needs a farrier to trim its hooves. (Donkeys don't generally need shoes from the farrier, unless you're going to be riding or driving in rocky country or on many paved roads.) Farriers are sometimes reluctant to work on donkeys. They don't usually see a lot of them, and the donkey's reputation for being difficult is widespread. Try to have your animals caught up and haltered before the farrier arrives. Provide a safe, covered place for him or her to work under, with a secure spot to tie the donkeys. Try to do your handling homework ahead of time so that the donkeys are quiet and well behaved when it comes to hoof work. If you pave the way for the farrier to do his or her job, you will have donkeys with healthy feet and less chance of going lame.

## Trailering Donkeys

At some point in a donkey's life, it will probably become necessary to trailer the animal somewhere (to the vet, to a show, to a new home). The time to teach your donkey to load is not five minutes before you're planning to leave. This is another job that ideally should be done during the

*Louise isn't sure she wants to load in the trailer. Being patient and having a helper urging the donkey from behind usually proves successful.*

Donkey Behavior and Handling **73**

# Getting a Handle on Things

Our experts discuss the behavior and handling of donkeys.

### Hooking the Good, the Bad, and the Newbies

"I have a technique for picking up feet that is more like what we do with our hands than using a rope. When we know that [the donkeys] understand what we want, we switch to a wooden cane with the half hook for a handle. Hold the straight end and run the hooked end down the leg, and then hook the ankle like you would with your hand and pull up. Then we step in if all seems well. No rope burns, no panic danger; [the cane] is not connected, so no foul if it does not go as planned—you just unhook and start again. [A cane also] puts you at a distance and at a different angle, maybe keeping knees safer; takes the direct pull off your back; and most important, your head is not leaned down and in range.

"I highly recommend this to anyone who trains, trims, or checks out critters with unknown attitudes. I learned [the technique] from an old fellow who used his walking cane. He had good critters, but his back just could not take the leaning down. It was more of a prompt for his [donkeys], but it sure works on the rogues and the newbies, too."

— Anne Hancock

### Eyeing the Gentle Ones

Sometimes fear, hormones, or something else will cause that sweet, loving pet to become difficult to manage and potentially dangerous. Size does not matter. Miniature jacks have been known to attack their handlers in the show ring, probably because of the excitement and stimulation caused by being around other jacks. But jennets also can be aggressive. It is not too uncommon for a jennet to threaten or even fight the owner attempting to work with a new foal."

—Joe Thomas

### Letting Go

"My rule of thumb when leading in a 'drag away' situation is that if I'm behind the shoulder and they're still going, drop the rope. I do it more to avoid rope burns than because of the danger of getting kicked, but the danger of getting kicked is very real. We were at the Altamont Fair one year, and a teenager was killed when the horse she was leading took off for the barn and she hung on to the end of the lead rope. The horse kicked out sideways and got her in the head, and it was fatal."

—Kris Anderson

donkey's young and malleable foalhood. If it hasn't, however, don't worry. It's not a terribly hard task to undertake with an adult. There are dozens of ways to go about loading a donkey.

Basically, it all comes down to whatever works, as long as the animal and the handler remain safe and the donkey isn't stressed more than necessary. Two of the most important hauling tips: donkeys load more easily in a stock-type trailer, and they are much happier being hauled loose rather than tied in the trailer.

To accomplish the loading, lead the donkey up to the trailer, and let the animal sniff and examine the doorway. After a moment, step up inside the trailer, and urge the donkey to do the same. Sometimes, the animal will surprise you by jumping right in without a second thought. More likely, it will remain in the doorway looking skeptical.

If you have a helper, this is the point at which he or she can assist by standing (out of kicking range) behind the donkey, urging it on by verbal means (clucking, making smooching noises, or clapping hands), or physical ones, such as tapping the donkey lightly with a whip. A white plastic bag tied to the end of the whip is often alarming enough to send the donkey hopping into the trailer in order to escape the scary rattling behind it.

If the donkey continues to resist loading, you can try a few other tricks. Occasionally, simple bribery will work, trading some treats or a bite of grain for compliance. The best method is to get the donkey close to the back of the trailer and tie the lead rope off on an inner ring or support inside the trailer. Then be patient for just a bit. Take slack out of the rope as soon as any forward progress is realized. It is very important not to let the donkey feel as if it can back up at all.

Chances are, the donkey will jump around, try to sit back, discover it is firmly anchored with no option available besides going forward, and ultimately leap in. With a smaller donkey, if its head is tied inside the trailer, two people can lock arms behind the donkey's rump and lift the animal in. If you're doing the job solo, however, just wait the donkey out; eventually the animal will load itself.

## Disciplining Miscreants

Donkeys are mostly mellow, easygoing critters, not prone to misbehavior. However, occasionally you may encounter an obstreperous youngster or a boundary-pushing jack that needs to be reminded who runs the show. Nipping is one of the most annoying bad behaviors. Young donkeys like to do it to get your attention, while jacks do it to see if they can dominate you.

Both should be soundly discouraged. A fist bumped quickly and firmly under the jaw, combined with a verbal "No!" will often surprise and confuse the miscreant. It usually doesn't see the fist coming but knows it's been caught in the act and chastised. Some folks pinch the upper lip and say "No!"

Whatever form of discipline you use, the rules are the same. You must do it quickly and decisively *at the time* the misbehavior occurs. You must be consistent—don't let the donkey get away with the behavior one time and punish it the next. You must be fair. Is the donkey truly misbehaving, or have you sent mixed signals about what is allowable and what isn't? Don't ever strike the donkey around the eyes or face, and don't ever "ear" the animal (twist an ear down as punishment or for restraint). Donkeys have extremely sensitive ears, and your donkey won't forget this painful treatment.

*Max reacts to the buffalo bull lying in the trail with a definite "ho-hum," merely taking the opportunity to snack. Because they are much less excitable than horses, Mammoth donkeys have become popular as trail mounts.*

## Saddling Up

The interest in riding donkeys is flourishing. Folks find donkeys to be much less stressful mounts; donkeys are far more laid back about riding than horses or mules. Many people have questions about using donkeys for riding purposes. Here are a few of the common queries:

**Q:** How much weight can a donkey carry?
**A:** It depends on the kind of terrain; the length of ride; and the donkey's age, build, and level of conditioning. A general rule of thumb is 25 percent of the animal's weight. This includes the weight of the tack along with that of the rider.

**Q:** What kind of bit and bridle does a donkey require?
**A:** No particular type. Most standard, 5-inch horse bits will fit a donkey. A full-cheek snaffle is a good choice for teaching directional control to a donkey that is untrained, but it doesn't have a lot of "stop." Some people prefer using a mullen mouth or correction-type bit for trained animals. Most horse-size bridle headstalls will work for donkeys, but the browband and throat-latch piece are often too tight and may need to be removed. (See Resources for suggestions on where to get equipment.)

**Q:** What kind of saddle is needed?
**A:** Again, this can vary considerably because donkeys come in all shapes and sizes. The best thing to do is to haul your donkey to a local tack shop, try various saddles on it, and have a knowledgeable person assist you in fitting. Otherwise, it's like buying a pair of shoes without trying them on—it will be hit or miss. Many donkey owners have had success using flex-tree or treeless styles, and some like the synthetic models. Usually, Mammoths will need the narrower, semi–quarter horse or regular bars, rather than full or quarter horse bars.

Virtually all of them will need an accompanying crupper or britchen to help keep the saddle from sliding forward.

**Q:** Do riding donkeys need shoes?
**A:** Sometimes. If the donkey is to be ridden for several consecutive days in extremely rocky areas, or doing a lot of road riding, you should err on the side of caution. It's better have the shoes and not need them, than need them and not have them.

**Q:** What's the best way to make my donkey start moving?
**A:** Donkeys aren't wired like horses or cars. You don't climb on, push the gas pedal, and take off. Donkeys don't respond as well to as horses do to round pen–type training, in which the animals are urged into forward motion by whips driving them in endless circles. They do, however, love to walk single file, following other animals. Therefore, early in the donkey's saddle training or the new owner/new donkey's relationship, go riding with other people who are mounted on quiet, gentle, and well-trained horses, donkeys, or mules. Following the other animals will help teach your donkey what is expected. Donkeys are very much "donkey see, donkey do." A donkey will be in its comfort zone when following other equines, and while doing this, the rider and donkey may fine-tune cues and reining skills, learn to cross formidable obstacles such as water, and work out their partnership. Eventually, the donkey will become comfortable enough to go on its own.

# Footloose and Fancy Free

Hauling donkeys is a different ball game than hauling horses. Most horses are led into a trailer and tied, standing either straight ahead or at an angle. Donkeys, however, suffer extreme balance difficulties and are fearful and agitated when forced to do the same. They will attempt to brace against anything they can and will remain frozen there for the entire trip. This often results in bloody, raw spots from continual rubbing against the solid object. If no wall is available to lean against, they will simply hang back against their halter and often end up with nasty sores in the throatlatch area.

Most donkeys are happiest riding loose inside the trailer. If given the choice, they will invariably ride facing backward; they seem much more sure of their balance doing so. They will also load much more readily into an open, stock-type trailer than into any other kind. With the smaller donkeys that can be more or less "hoisted" in, this is not a big issue. However, Mammoth donkeys that choose not to cooperate in the loading process can be a challenge.

# chapter SIX

# Donkeys in Sickness and in Health

Generally hardy, sturdy animals, donkeys, with good husbandry, should have few health problems—good husbandry being key. As is true with most things, when it comes to keeping donkeys healthy, an ounce of prevention is worth a pound of cure. That means taking proper care of your donkeys on a daily, monthly, and yearly basis. Making sure they get the vaccinations and other preventative care they need.

Key as well to good health is being able to recognize what is and isn't "normal" for your particular donkeys. This knowledge will proved to be crucial when it comes to staving off serious illness and can assist your veterinarian in making a diagnosis. It is also important to know what kinds of parasites and illnesses are most likely affect your donkeys.

## Keeping Your Donkey Healthy

A healthy donkey should be bright and interested in everything. Its summer coat should be shiny and not dull, rough, or patchy. Although the donkey's winter coat may look a bit "rougher," especially close to spring when the dead hair is ready to be shed, the coat should be free of obvious signs of neglect. The animal's eyes and nose should be free of discharge. It should stand squarely on all four feet, with no heat or swelling in the limbs. The donkey should clean up its feed and be eager for more. Its manure should be formed into small, moist balls, varying in color depending on feed, but solid in consistency and not containing pieces of undigested grain.

Maintaining your donkeys' health means using common sense in your practices. If possible, quarantine newcomers to the farm for a week or two; if not, at least limit nose-to-nose contact. Keep good records. Written records of vaccinations; deworming, breeding, and foaling dates; illnesses; and any other medical incidentals can be invaluable. Stay up to date on vaccinations, follow a deworming program suitable

*These jennies on the Leon Oliver farm, in Cornersville, Tennessee, are in their thirties. With good care, donkeys can live to great ages.*

for your region, have teeth checked and attended to at least once a year, and have a farrier regularly perform hoof work.

## Vaccinations

Vaccinations are an important part of any donkey's health program. You may find it difficult, though, to decide what to vaccinate for and when. Incidence of disease differs considerably from region to region, safety and efficacy of the vaccinations vary, as do personal circumstances. Sometimes drug companies and media blow risks out of proportion, never really questioning the safety and necessity of the vaccines. A person who never leaves the farm with his or her animals will not have the same vaccination needs as one who regularly shows, trail rides, or trades donkeys.

### Basics of Vaccinations

Basically, owners need to consult with the vet and vaccinate for the diseases affecting their regions (or the regions they will be traveling to) and for which there exist safe, effective vaccines. In general, equines that do not travel should have, at a minimum, the annual vaccines for eastern/western encephalomyelitis, influenza, tetanus, and West Nile. Other vaccines your vet may suggest include those for rhinopneumonitis (particularly for pregnant jennies), strangles, rabies, and Potomac fever.

Although a veterinarian should definitely be consulted about vaccinations, many owners can learn to give their own shots. If you decide to do so, be sure to buy vaccines from a source you trust to have handled them properly, kept them refrigerated, and not allowed them to become outdated. These sources can include your vet, farm supply stores, and mail-order equine catalogs.

## Administering Vaccinations

You should read labels first to confirm the procedure. In most cases, vaccinations are administered intramuscularly (IM). Begin by securing the donkey safely. Then select a clean, dry spot in the muscular area of the side of the donkey's neck for injection. People differ on whether it's necessary to swab the injection site with alcohol before administering the vaccine. Doing so can't hurt.

If you purchased the vaccines individually, they will come prepackaged in syringes and ready to use. If you purchased a vial of vaccine, then you will also need to purchase 3-cc syringes and 1-inch or 1½-inch needles (20 or 22 gauge). Use each syringe and needle just once. When

*A preferred location for giving a donkey intramuscular injections of vaccine and medication is outlined here on Lolli's neck.*

the vaccine is in the syringe and ready to use, unscrew the needle and separate it from the syringe. Tap the donkey's neck a few times (this is a distraction technique), then pop the needle in quickly and smoothly at a 90 degree angle, deep into the muscle, all the way up to the hub of the needle. Carefully reattach the syringe to the inserted needle. Pull back a bit on the plunger to make sure you didn't send the needle into a blood vessel.

If this is the case, blood will appear in the syringe, and you need to pull the needle out and start over at another spot. If not, firmly depress the plunger to administer the vaccine, and massage the spot for a few seconds after withdrawing the needle. Most donkeys will barely flinch, but a few may jump when the needle goes in. Be ready for this; don't get stepped on or knocked down. Safely dispose of all used syringes and needles.

Although it is rare, animals do sometimes have anaphylactic reactions to being vaccinated. Discuss this possibility ahead of time with your vet, and ask for a dose of the prescription drug epinephrine to keep on hand to treat this potential emergency. The donkey may experience a bit of lethargy or soreness at the injection site, as well as have a small bump there for a day or two after vaccinations. These usually clear up without the need for any treatment, but if you feel concerned about the degree of reaction, consult your vet.

## Dental Care

Just like us, donkeys benefit from a regular visit with the dentist. Not only can dental problems cause digestive upsets from improper chewing of feed, but they can also contribute to an overall thin, unthrifty appearance. Young stock often need wolf teeth and caps removed, while older donkeys frequently develop sharp points on their teeth, which require filing ("floating") to smooth their edges. Your veterinarian will provide these services or perhaps recommend a local, trained dental technician.

## Farrier Care

As discussed in the previous chapter, farrier care is a very important part of good donkey husbandry. Poor hoof care can cause or exacerbate illnesses such as laminitis (founder), hoof abscesses, and thrush, a bacteria that thrives in dark, moist areas of the hoof, especially when it is packed with mud or manure. Thrush

# Castration in Donkeys

One very important difference between donkeys and horses is the method used during castration. Vets castrating horses commonly just crimp blood vessels with an emasculator, which works fine for these equines. Donkeys have much larger vessels than horses do and corresponding blood supply to the area. Your vet must be educated regarding that difference and not only crimp but also ligate the vessels. Veterinarians will sometimes be reluctant to do this because of a slightly increased risk of infection, but as one owner put it, "You can cure an infection, but you can't cure dead." Horror stories abound of donkeys bleeding to death following castrations. Make sure your vet proceeds with all due caution and ligates those vessels.

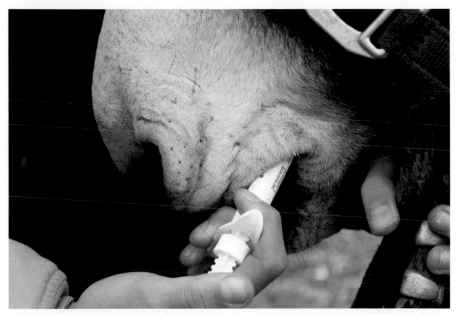

*The dewormer paste is deposited on the back of Lolli's tongue. It's a good idea to hold the head elevated with the mouth closed for a few moments to make sure the donkey swallows the contents.*

can eat away at the frog and sole of the hoof if left untreated. Finding and using a good farrier will go a long way toward keeping your donkey's feet healthy.

## Parasites

Worms, lice, ticks, flies, and mites— these are all parasites donkeys are subject to and must be treated for, internally and externally.

## Internal Parasites

Like other equines, donkeys are subject to infestations of internal parasites, commonly referred to as worms. These worms live in the stomach and intestines, their eggs having been ingested by the donkey when it grazed on pastures with contaminated droppings from other afflicted equines. Common symptoms of heavy infestations include thin, unthrifty appearance, potbelly, rough coat, and occasionally a cough.

## Deworming Programs

Foals should be started on a deworming program by eight weeks of age. Consult with your local veterinarian for recommendations on frequency of deworming. This can vary depending on region, on the number of animals you house and on how much acreage, and on the age of the donkeys. Most folks deworm from two to six times a year (some vets recommend deworming foals more frequently), rotating between the different classes of anthelmintics (dewormers) to assure all types of parasites, in all stages of development, are eradicated. These classes include ivermectin, pyrantel pamoate, febendazole, moxidectin, and praziquantel. Your vet can give you advice on which class to use and when.

Moxidectin doesn't have the margin for error that the other dewormers do, which makes it tricky to use. You need to have a precise knowledge of the donkey's

*A Miniature foal goes after an annoying itch with her teeth. If a donkey is doing a lot of chewing and rubbing, an examination of the skin is in order.*

weight to administer the drug correctly. Because equine weight tapes are not as accurate when used on donkeys as on horses and most owners don't have access to a large-animal scale, many of them forgo the use of moxidectin.

## Administering Dewormers

Dewormers themselves come in easy-to-use tubes, scored by weight. Simply "dial" the ring on the tube's plunger to correspond with the weight of your donkey, insert the tube into the side of the donkey's mouth on the back of its tongue, and depress the plunger. It's a good idea to use your other hand to hold the donkey's head up and to massage his throat immediately after you administer the dewormer. Sometimes donkeys can be quite crafty about spitting out a big white glob of wormer long after you think it's gone down the hatch.

*A word of warning*: If you acquire a donkey you think is extraordinarily infested with worms, proceed cautiously. A

### DID YOU **KNOW**

Donkeys are the natural host of parasites called lungworms, which usually cause donkeys themselves no problems. Lungworms can be passed in the donkeys' manure, however, and from there infect any horses using the same pasture. For this reason, it is advisable to treat donkeys with an ivermectin product before pasturing them with horses.

full dose of dewormer can cause a kill off that is too heavy. This can lead to impaction colic due to scores of dead worms balling up in the intestines. Again, it is prudent to get your vet's recommendations on what deworming product to use and how much. It is also a good idea to have your vet run a fecal test on your donkey (or a few sample animals if you have a herd) at least once a year. This can reveal what parasites your animals are infested with and indicate the effectiveness of your deworming.

In addition to using anthelmintics, you can help lower the parasite infestation of your donkeys by keeping manure removed to an area they can't access. In addition, mow and harrow pastures (rotating animals on and off helps, too). Finally, always use feeders; it is usually not safe to feed directly off the ground.

## External Parasites

External parasites (flies, lice, ticks, and mites) are often the bane of a donkey's existence. When living in areas with enough moisture to encourage a large insect population, the desert-bred donkey suffers mightily.

### Lice

The donkey's coat grows long and shaggy in the winter and is slow to shed out in the spring. This sets up a perfect breeding ground for lice (both sucking and chewing varieties), which tend to afflict donkeys in late winter and early spring. If you notice your animals rubbing out bare spots around the tail head and along the mane area during this time of year, suspect lice, little brownish creatures about the size of fleas. Treating a donkey with ivermectin dewormer should help kill off

# First Aid Kit

Here are some general guidelines on what medications, instruments, and other items to keep in a first aid kit for your donkeys:

- Analgesics/anti-inflammatories, such as Banamine/phenylbutazone ("bute")
- Antibacterial wound ointment, such as Furazone
- Antiseptic ointment, such as Nolvasan
- Gentle iodine—for treating newborn foal navels, abscesses, fungal infections, and wounds
- Hydrogen peroxide
- Latex gloves
- Lubricant jelly
- Ophthalmic (eye) ointment
- Rectal thermometer
- Scissors
- Sterile gauze roll and self-sticking bandages, such as Vet Wrap
- Stethoscope—to check heart rate. Normal rate is 30 to 40 beats per minute; its higher for a nursing jenny and higher still for a foal (60 to 80). Listen to the heart on the chest's left side, just behind the left elbow.
- Twitch (restraint device)

*Dixie, a Mammoth, models a fly mask. As the name implies, this device keeps flies, as well as dirt and other irritants, out of an animal's eyes. Donkeys can see through the mesh mask.*

sucking lice, and dusting the donkey well with a powder made for this purpose will eradicate both types of lice. If the weather is warm enough, body clipping and bathing your donkey with a pyrethrin-based shampoo is also effective.

## Ticks

Ticks come along right on the heels of lice, in late spring and early summer. They are especially bothersome in pastures with tall grass and in wooded areas. Keeping pasture grass mowed and keeping donkeys out of wooded areas may help control the pests to some degree, but it's hard to keep donkeys entirely tick free. Various brands of fly sprays are also labeled as being effective for ticks. Some of the spot-on-type products (those applied directly to the skin) designed to be absorbed into the bloodstream are useful against ticks.

## Flies

Just about the time that the lice and ticks finish tormenting the donkeys, the flies take their turn. For some reason, when donkeys are in the process of shedding the winter hair off their legs, the flies go wild for them. Leg of donkey must be a major fly delicacy. Flies will chew a donkey's legs into bloody, raw messes. This situation requires full-scale war against the insect enemy, which can include an arsenal of fly spray, spot-ons, leg wraps, natural remedies such as apple cider vinegar and garlic, fly predators, and fly traps and strips.

The most success we've had in dealing with flies has been by keeping manure to a minimum, using fans in the barn, applying fly spray as needed, and coating the donkey's legs with a repellent/wound ointment such as SWAT during

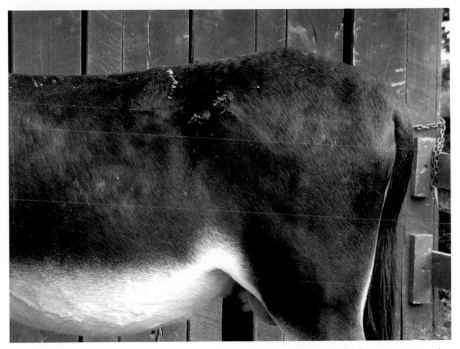

*Summertime often produces sores such as the one on this Mammoth's hipbone. It is caused by the donkey's swinging its head around to knock off flies and hitting the area with its teeth.*

the worst of the fly season. Sometimes, when the flies are swarming around the eyes of our donkeys, we also put mesh face masks on them.

During late summer and early fall, you might see your donkey suddenly throw its head up and sprint wildly for the barn. Closer inspection will likely reveal a small, beelike critter attempting to land on the donkey's legs to lay its yellow nits (eggs). This is the bot fly, whose larvae migrate to the donkey's stomach and cause multiple health problems. Fly spray, sandpaper-type tools, and razors can all be used to remove the nits. Ivermectin should be used in early winter to break up the bot cycle.

### Mites

Mites are tiny creatures that feed on the donkey's surface skin or burrow into it. They cause excessive itching, and an infestation will often make a donkey rub itself raw. Mite infestations occur much less often than infestations of lice. To find out whether mites are the problem, your vet will take a skin scraping to examine under a microscope. He or she will advise you on the best form of treatment, which will probably include enough medication for your entire herd of donkeys, since, as is also true of lice, mites are spread from donkey to donkey, and if one donkey is infested, all are likely to be infested.

## Treating Common Conditions

The following are a few health-related problems you need to be aware of so you can look for the signs and take the appropriate actions. Most of these are common and are not life-threatening conditions.

## Hoof Abscesses and White Line Disease

Eventually, most equine owners will have to deal with a hoof abscess. For whatever reason (maybe the all-too-common neglect and the infrequent trimming and cleaning), donkeys seem to have more than their share of abscesses. Feet allowed to grow too long are prone to bruising in the toe. An abscess begins when something makes a break in the bottom of the hoof that allows bacteria and dirt access. Puncture wounds, deep bruises, and nail holes from shoes can all be culprits.

Most abscesses develop over a couple of days, with the donkey becoming progressively lamer. The animal often spends a lot of time lying down. Your farrier may be of more assistance than the vet in this situation. Depending on the location of the abscess, as well as other variables, the farrier may be able to locate and pare out the abscess, providing almost instant relief for the donkey.

Follow-up treatment may involve soaking the affected foot in Epsom salts, packing the pared-out abscess with cotton balls soaked with iodine, and wrapping the hoof with Vet Wrap or a similar foot bandage to keep the packing in place. The foot needs to be re-treated daily as long as pus drains from the abscess. Make sure the donkey is up to date on tetanus vaccinations, as the animal is especially vulnerable to tetanus following a hoof abscess.

White line disease (WLD), or "seedy toe," is not really a disease but an infection or condition of the hoof wall. It is caused by various types of bacteria and fungi that infiltrate and destroy the wall's keratin tissue. One of the most recognizable characteristics of the disease is a soft, crumbly, white substance, easily scraped out, between the hoof wall and the white line on the bottom of the hoof. When you examine the underside of the hoof, you will see the white line as a light-colored, thin groove that connects the outer wall to the sole. The affected hoof may have a hollow sound when tapped by a hammer.

White line problems can be the result of previous bouts with abscesses or founder. Moisture seems to play a big role in the problem, as the condition is not nearly as common in drier regions. Treatment can be frustrating, and you may have to work closely with your farrier to eradicate WLD in your herd. Promoting good hoof growth, keeping the environment clean and dry, cleaning out the affected feet, and treating several times a week with iodine or other fungicides will all go a long way toward successfully dealing with this condition.

## Spring Skin Grunge

Every spring, many owners are appalled to notice how grungy their donkeys have gotten. Because it grows such a thick coat and wears that coat nearly year-round, the donkey is susceptible to skin problems. Moisture gets trapped in its long, hairy coat and mixes with the dirt the donkey loves to roll in, denying the skin access to sunlight or air—the perfect recipe for skin grunge. This type of coat is also an ideal habitat for nasty critters such as lice, so by late winter and early spring, the donkey is an uncomfortable, itchy, ratty-looking character, at least in parts of the country where cold weather and moisture are most prevalent.

The animal may be suffering from ringworm (not a worm at all, but a fungal condition characterized by round lesions with a raised, crusty outer edge). Rain rot (another fungal infection), usually located along the donkey's back, is

*In late spring, Matilda's coat appears mangy and rough as patches of the dead winter coat get rubbed out. The shedding process is a slow one, often finishing only shortly before a new winter coat begins to grow in.*

common. Scurfy patches of hair that you can easily pull out in small clumps are signs of rain rot.

Unfortunately, many of these conditions will prove difficult to eradicate completely until the underlying cause (the old, heavy, dead hair coat) is removed. As early as is prudent, the donkey with skin grunge should be body clipped and bathed with iodine or some sort of antifungal shampoo. This, combined with sunlight and air gaining access to the skin, will usually be enough to restore the donkey's coat to good health. It is difficult to prevent these problems, but good health and hygiene will lessen their impact. Keeping the donkeys as clean as possible, frequent brushing, adequate shelter from moisture, and application of pesticides are all methods of reducing the problem of skin grunge.

## Sarcoids

Sarcoids are nasty-looking (but noncancerous) growths most commonly found around the head and the ears and occasionally on the body or lower limbs of the donkey. No one really knows why donkeys are so prone to getting them, and there is

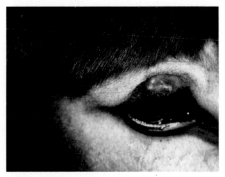

*This poor jenny has a sarcoid tumor on her eyelid. These tumors, though common and benign, respond best when treatment is initiated as soon as they're noticed.*

no way to prevent them. Various treatments may be tried, including surgical removal, banding, freezing, or application of herbal salves. Once removed, sarcoids may be gone for good, but they have a fairly high incidence of recurrence.

## Umbilical Infections

Mammoths, in particular, suffer this problem. The navel area of young foals can become infected, and if not treated, the infection can quickly lead to death. All newborns should have the umbilical stump sprayed or dipped in gentle iodine once or twice a day for the first few days of life.

After that, keep an eye on the area for the foal's first month. If the stump becomes hot, swollen, and weepy and the area is tender to palpation, or if the foal seems lethargic and does not nurse well (you can tell this by the engorged bag of the dam), take the foal's temperature ASAP. Temperatures over 102 degrees, combined with the other symptoms, indicate an umbilical infection, and a vet should be summoned at once. Babies can crash with alarming swiftness and need to be started on antibiotics quickly.

Keeping foaling jennies in clean surroundings and treating foals' navels with iodine can help reduce the incidence of infections, but sometimes, in spite of owners who do all the right things, the infection will still occur.

## Jack Sores

A problem most commonly associated with Mammoth donkeys, jack sores are one of the more mysterious longear ailments. Vets often identify them as summer sores (cutaneous habronemiasis), which are caused by the larvae of stomach worms. Some researchers, however, doubt that the syndrome affecting donkeys is exactly the same, though symptoms and treatment are similar. The most common site for the jack sore is the inner hock of the donkey's hind leg, presenting as a horizontal laceration right across the "bend." Crusty and weepy, the sores may remain for weeks or even months and usually are worse during the warmer summer weather.

Treatment is usually a combination of ivermectin (to kill the worm larvae) and a fly repellent or wound ointment. One recipe calls for 4 ounces of 10 percent pyrethrin and 8 ounces of mineral oil, to be mixed in a one-gallon garden sprayer and then sprayed on the lesions. Cauterization or surgical removal of granulated tissue may be required in severe cases, but in some cases, spontaneous healing occurs.

What is mysterious about this syndrome is that it seldom affects anything but jacks and is much more prevalent in Mammoths than in smaller donkeys. It has been around as long as there have been Mammoth Jackstock in the United States. Old-timers had many theories about jack sores. They generally believed feeding too much "hot," or rich, grain caused them.

## Upward Fixation of the Patella

The stifle joint on the hind leg of a donkey is similar to our knee. It has three ligaments running to the outside, middle, and inside of the leg. Occasionally, during flexing, the ligament running along the inside of the leg gets caught on a notch on the end of the thigh bone, causing the leg to become completely "locked up" in an extended position, dragging straight out behind the donkey. Usually, the locking is very brief and results in a jerking or hitching upward motion as the ligament snaps back into place. This condition is most commonly seen in young, growing animals, and for some reason it seems to occur more often during cold, wet weather.

*This Mammoth jenny watching over her foal has a jack sore on her right hind leg. It's unusual for a female donkey to develop one of these mysterious sores, which are most frequently found on male Mammoths.*

Although alarming to witness, these episodes do not cause the donkey distress. Most cases are resolved with time, and exercise can help by tightening stifle ligaments and strengthening quadriceps. If the condition persists into adulthood, a minor surgery can be performed to cut the medial patellar ligament.

## Treating Serious Illnesses

As stated earlier, when it comes to donkeys, few things constitute a medical emergency; the following four conditions are exceptions. These are the illnesses that you don't want to try to deal with yourself—a delay in acquiring veterinary assistance may mean the difference between life and death.

## Colic

When it comes to equines, colic is another way of saying bellyache. Unlike humans and many other species, equines cannot vomit, so when they have gastrointestinal distress, their bodies have few ways to quickly relieve it. Colic can be caused by many things, including a heavy intestinal parasite load; sudden changes in and/or overconsumption of feed; consumption of spoiled, moldy feed; dehydration; and stress. In horses, colic symptoms include excessive

# Signs of Illness and What to Do

With good husbandry, donkeys—generally healthy and vigorous—seldom become ill. However, because donkeys are stoic creatures, it does pay to make note of any changes in behavior or appetite and to thoroughly investigate their causes. A donkey backing away from its feed is a red flag. If you have any concerns about whether you have a sick donkey, you can perform a few tests.

First, take a rectal temperature, which should be around 99–100 degrees. Then listen to gut sounds (press your ear to the donkey's belly on each side in the flank area). There should be plenty of gurgling and rumbling sounds. Silence, never good, can indicate a shutdown of the intestines.

*A normally sociable donkey stands off by herself in a field, unusual behavior that could indicate illness or some other serious problem. Any change in a donkey's behavior is cause for concern and should be investigated.*

Take a pinch of skin along the side of the neck. If the skin flattens in one second or less when released, the donkey is normally hydrated. If it stays pinched up and flattens slowly, dehydration should be suspected. Now press your thumbs into the donkey's gums to check capillary refill time. Color should flood back into the gums in two seconds, and the gums should be moist and pink. Pale or bright red, grayish blue, or yellow gums can be signs of illness. Make all of these assessments and report them to the vet when consulting him or her about your concerns.

In addition to the above, here is a list of symptoms that indicate treatment could be urgently needed:

- Loss of appetite
- Profuse bleeding that can't be stopped
- Suspected bone fracture
- Wound or cut deep enough to require stitches
- Eye injuries of any type, including sudden cloudiness, puffiness, swelling around the eyes
- Temperature over 102 degrees
- Inability to stand or incoordination while walking
- Swelling of any body parts, or body parts that are hot to the touch
- Straining to urinate or defecate

rolling, pawing, getting up and down and looking back at the sides, kicking at the belly, sweating, and refusing to eat.

In the more stoic donkey, the symptoms are considerably more subtle. The donkey will probably seem depressed and uninterested in feed, lie down frequently, and stay away from its fellow donkeys. Checking for gut sounds may reveal an ominous silence from within.

This is a situation that many vets mistake for "mild colic." They often suggest treating with an anti-inflammatory/analgesic such as Banamine or merely "keeping an eye" on the animal. Be polite, but insist that a vet personally examine the donkey. Our experience has included a donkey that showed little in the way of overt symptoms—even her vital signs were barely elevated. But an ultrasound revealed grossly distended intestines close to rupturing. She had to be rushed in for emergency surgery.

To avoid cases of colic, feed good, clean hay and grain; have plenty of fresh water available at all times; make any changes to diet slowly; keep up a good deworming program; and limit stress as much as possible.

## Founder

Founder (properly known as laminitis) is an inflammation of the layers (or laminae) of the hoof. The classic stance of foundered donkeys is stiff legged and rocked back on the heels, with front feet propped out ahead. During an acute attack, a donkey will have heat in the hooves and a rapid pulse at the coronary band. In extreme cases, the coffin bone in the hoof separates from the laminae and begins to sink, causing severe pain and lameness. In serious chronic cases (such as that of racehorse Secretariat), the animal must be euthanized.

Founder can have many causes, not all of them clearly defined, but one of the more

*Eeyore displays the beginnings of "fat pones" along the back. When a donkey gets these, it is headed for obesity and the potential for many health problems. Diet time!*

# Health Issues

Our experts sound off on various donkey health issues.

### Keeping Worms at Bay

"I have begun checking fecals, rather than worming on a six- or eight-week schedule. This way, there is no resistance built up to any wormer in the donkeys' systems, and I don't give unnecessary chemicals to [the donkeys]. I do make a point of making sure once a year they are wormed with a product that will take care of tapeworms, as they do not show up on fecal samples."
— Ann Firestone

### Giving the Gift of Knowledge

"I give our vets a membership in the American Donkey and Mule Society so they get *The Brayer* magazine, which has some technical health articles and also gets the new vets and vet techs who haven't had much contact with donkeys interested in them."
— Alison Kern

### Being Aware of Issues

"There are two health issues associated with donkeys that I think people should be aware of. If these are recognized and treated properly, [owners] can avoid a lengthy illness and possible death. The first is hyperlipidemia. The best published information that I have found is in the *Australian Veterinary Journal*, volume 76. The second is guttural pouch disorders. The best published information can be found at www.ivis.org under 'Respiratory diseases.'"
— Deby Cochran

### Bugging Off!

"Here we are in muggy, hot, Oklahoma, already battling the pest onslaught. Very early this spring, we purchased an EPPS fly trap and put it up before the first sign of flies. It seems to work really well, traps a lot of flies, with only water and dish soap needed. We put up the little yellow, sticky traps everywhere we can find space. We are currently awaiting our fourth shipment of fly predators and have a manure spreader on order. As far as the 'kids' themselves, we are currently using Farnum's WIPE, and we are very pleased with this product. Also, we are using sprays, so depending on who is available we 'team de-fly.' One of us uses a rag and the Farnum's on faces, ears, and legs (especially around the hooves), and the other sprays the big parts."
— JoAnnie Kale

### Ligating Those Jacks

"I had my jack gelded when he was sixteen, and it did not go well because the vet wanted to do it his way instead of ligating the vessels as I suggested. Ligating the blood vessels is especially critical with older jacks, as the vessels develop so much connective tissue that the walls are rigid, so crushing them does just about nothing. So the vet left the clamps on and then wanted me to take them off after six hours, which I did, and it was like turning on

common ones is overeating, particularly on lush spring pastures or from an "accidental" break-in to a grain storage area. Stress can also be a big contributing factor, and obese donkeys are always at a higher risk. If you suspect laminitis, consult the vet immediately. The earlier treatment begins, the better the outlook.

## Hyperlipidemia

Elevated level of fat in the blood cause this condition. It is especially prevalent in pregnant or lactating jennies. Obesity combined with stress can bring it on. The donkey's appetite shuts down. Other symptoms are depression, uncoordinated movements, and muscle twitching. The vet will pull a blood sample; high readings of triglycerides and cholesterol will help solidify the diagnosis. Treatment must begin immediately, but even so, the prognosis is often poor. Hyperlipidemia carries a 60–85 percent mortality rate. Establishing a high-carbohydrate, low-fat diet is important, but the main thing is to just get the donkey to resume eating. Intravenous glucose may be administered. You may also try beet pulp soaked in fruit juice, apples, carrots, or freshly picked grass to tempt the palate of the patient. The most important step you as an owner can take to prevent this disease is to keep your donkeys from becoming obese.

## Respiratory Distress

Poor nutrition, heavy worm infestation, and exposure to cold and rain can predispose donkeys to pneumonia, which they can also develop as a secondary bacterial infection following a bout with influenza. Rapid, labored breathing, often combined with a loss of appetite and depression, as well as congestion and rattling or wheezing sounds, should be cause for an urgent call to your vet.

two faucets, the blood ran out of him so fast. So we had to hobble him and get the clamps back on before he bled out, then we left them on for another twelve hours, and this time only took one off.

"He was still bleeding profusely, so we got the clamp back on and left them on for another twelve hours. The first time I put the clamps back on, I didn't have any sterile gloves, so of course he got infected. That led to three weeks of penicillin injections, oral antibiotics, bute, hosing, and twice-a-day exercise. He ended up foundering but did survive, although I think it was probably a very close call.

"My advice is don't have one gelded unless you find a vet who understands that the blood vessels have to be ligated. Hundreds, and probably thousands, of jacks have been ligated when castrated. While it does increase the risk of infection, most don't get infected."

—Kris Anderson

# chapter SEVEN

# Breeding Donkeys

**F**uzzy, affectionate, and endearing, donkey foals are some of the most appealing creatures on Earth. Many folks are tempted to breed their jennies so they can experience the enjoyment of having a foal. No one, however, should go into breeding without clearheaded goals and plans. The recent economic downturn has severely affected the equine industry.

That being said, a good market always exists for the very best animals. If a breeder develops specific goals and produces what is in demand, he or she will find eager buyers. The market for, registered stock is usually better than it is for grade stock. Miniatures seem to have their own niche market, particularly the well-bred, well-conformed show stock in fancy or unusual colors. You will find the demand for saddle-trained Mammoth Jackstock to be exceptionally high. Generally, with the possible exception of colorful spotted animals, standard donkeys are the least sought-after donkeys. There is such a surplus of feral burros available for adoption from the Bureau of Land Management that few people specifically breed that size donkey.

Regardless of what type of donkey you are breeding or what your goals for the breeding program may be, your choice of breeding stock should include the very best animals you can possibly afford. The old saying "Quality doesn't cost, it pays," rings as true for livestock breeding as it does for anything else.

## Selecting Breeding Stock

Three main criteria in selecting breeding stock should be conformation, size, and disposition. Prospective breeders must do their homework and educate themselves on these criteria, especially conformation. Talk to other breeders, attend shows and symposiums, go to sales, and join clubs. Subscribe to all industry publications. The more educated the breeder, the more successful he or she is likely to be.

*Eeyore comes to check out what's going on. Look for a bright, interested expression when you shop for a donkey for breeding purposes; disposition is important.*

Basically, good conformation includes good overall balance. All of the donkey's parts should seem to fit smoothly together, giving a pleasing visual impression. Legs should be straight, with no knuckling over at the knees or crouching up under the body behind. A donkey's knees (particularly a Mammoth's) can appear a bit rougher and knobbier than those of a horse, but all animals should move smoothly and show no signs of joint impairment.

The back should be straight and level, although an older brood jenny may be forgiven a slight sag when it is the result of multiple pregnancies. The head should be attractive, with long, well-set-on ears (that is, set directly on top of the head, with little distance between them). Do not purchase a donkey with lop ears, swiveling out on the sides of the head and drooping.

All breeding stock must have good bites. This means that the teeth must meet evenly (no malocclusion, or bad bite). When you shop for prospective breeding stock, this should be one of the first traits that you check for; any donkeys that have under- or overbites should be culled from consideration.

Another important point to check, if you're looking at jacks, is whether the animal has two testicles. If you cannot see or at least feel both of them, he may be a cryptorchid (a jack with one undescended testicle.) This condition and malocclusion are inherited defects and undesirable traits in a breeding animal.

As for size, if you want to breed Miniatures, look for stock that will stay below the maximum height requirements. If you want to breed Mammoths, look for stock that will mature to that height. Whatever breed you choose, make sure the animals have a good disposition (are friendly and easily handled).

Find out as much as possible about the jenny's prior production and about her immediate family. Jennies from fertile, good-producing bloodlines will be your best bet. If you buy a jack as your herd sire, he will be the cornerstone, so choose well. Good looks and good conformation do not mean a jack will be a good producer. If you're purchasing a mature jack, look at as many of his offspring as you can.

## Breeding

Donkeys usually cycle year-round, although sometimes not as regularly during the hottest days of summer and the coldest days of winter. In the Midwest, late spring and early fall seem to be the best seasons for foaling, but breeders can pretty much choose any time of the year that they desire.

Jacks may be sexually mature by eighteen months of age (or even younger for minis), but it's best to allow your jack more time to mature both mentally and physically before embarking on his breeding career. Before you begin using him for breeding, teach him manners and get him thoroughly accustomed to being handled all over, led, tied, and so on. It will make him much easier to deal with when the hormones kick in.

# Keeping a Jack—Worth the Hassle?

Unless you have more than three or four jennies, you will probably be better off taking your jennies to other farms and paying a stud fee to have them bred than keeping your own jack at home. Jacks have extra needs when it comes to

fencing and housing as well as handling. They should never be seen as pets, and unsupervised children should not be allowed near them. Because a jack is normally friendly, it's very easy for us to forget he is a stallion, possessed of all the testosterone and instincts of one. Although a jack is seldom aggressive toward people, when one strikes, he does so with lightning quickness and next to no warning. Jacks should be handled with firmness and respect by those experienced in dealing with them.

*Colonel surveys his kingdom from his separate quarters on the farm. Think twice—or thrice— before purchasing a jack for breeding; jacks must be handled with care and require special housing.*

*Mammoth donkey Rebel displays the flehmen response. This behavior is typical in jacks sampling the air for pheromones of a jenny in heat.*

Jennies may start cycling as young as six months of age but should not be bred before three or four years old. This a good reason to separate the sexes by weaning age. Although jennies are sometimes bred at two years, doing so can compromise their future growth, and a jenny is not mentally ready at three to mother a foal. There should be no great rush to put a jenny into production; with good care she can remain productive into her twenties.

The jenny will usually signal the beginning of her heat cycle with some unmistakable signs. She will make vigorous chewing or jawing motions, sometimes drooling, with her ears laid back and neck outstretched; simultaneously, she will squat to pee. The jack apparently find this attractive picture irresistible; he comes running, honking with every step. In a herd, jennies will often show heat to each other, chewing and mounting other jennies. If no other equines are present, your jenny may not be as obvious about being in estrus. You may have to lead her up to a jack's pen or pasture (on the other side of the fence) to see if she shows signs of being in season. Never put a jenny in with a jack unless you are certain that she's in heat; jacks can be extremely aggressive, chasing, biting, and knocking down jennies.

The jenny generally remains in estrus for three to eight days, with the cycle ending when ovulation occurs. If she is not bred, or fails to get pregnant, she will cycle again in approximately twenty-one days.

There are several methods employed as far as the actual breeding process goes: in-hand breeding, pasture breeding, and artificial insemination. Which you choose depends on several factors.

## In-Hand Breeding

In this breeding scenario, the jenny is tied up or otherwise restrained, while the jack is led to her for serving. He is usually equipped with either a breeding bridle (a headstall with a bit in his mouth) or a stud chain (a length of chain, 18 inches or so, attached to the lead rope on one end, run through the halter rings under the jack's chin, and snapped to the ring on the right side of the jack's halter.) This gives the handler a little added control and leverage.

The jack may be led up to sniff around the jenny's hindquarters and acquaint himself with her alluring scent but is usually not allowed to mount until he is fully "dropped out" and erect. The nonerect jack often enjoys leaping on the jenny and biting her neck and shoulders rather forcefully. The jack may like this type of foreplay, but the jenny seldom appears to do so. After a period of time, which can vary from a few minutes to a few hours, the jack will become erect and serve the jenny.

Mammoth jacks, in particular, can be maddeningly slow breeders and seem to take eons to achieve an erection and do the deed. You will find that it helps to stick with a very strict routine. Have as few distractions around as possible. Don't ever have a

*A stud chain runs under the chin of this jack and snaps to the ring on the right side of his halter. The chain gives a handler greater control over the jack during an in-hand breeding.*

# Donkey Breeding

Our experts offer some advice on donkey-breeding programs,
on pregnancy, and on foaling.

### Have a Dream

"Have a dream, a goal, and a plan before undertaking a breeding program. Be aware that large, drafty Mammoth jacks can have a very low sex drive and take a long time to get ready. Choose the jack for your jenny very carefully. It is a long 12 months, and you want to be pleased with the foal you get."

—Bobbi Ward

### Analyze Your Program

"As a breeder, keep your goals in mind. Breed for quality, not quantity. You need to analyze your breeding goals each year and decide if they need to change to match the current trends or if you even need to breed your jack to jennets during that year."

—Lynn McMillan

### Check for Twins

"One bit of advice as a new breeder I have to offer is the importance of ultrasound. Of four jennies I had ultrasounded, two had twins. A twin pregnancy reduction is $200 and up [where I live] but well worth the other alternative of having the twins. Having a competent veterinarian is crucial to any breeding program."

—Sue Elliot

### Don't Squeeze Too Much

"People nervously anticipate the arrival of the young foal, fearing something will go wrong. The worry starts with the first signs of swelling of the udder. This usually starts four to six weeks before foaling. It is common for the udder to swell, then go down, then swell a number of times. As the jennet gets close to

foaling, the udder will usually stay enlarged but does not get really tight until close to foaling time. Then the teats get engorged and may leak milk. This often leaves a crusty substance on the end of the teat. This is called waxing and indicates foaling is close. I squeeze one to two drops in my hand daily to check for color. Once the milk turns white, foaling usually occurs within twenty-four hours. Anxious owners need to be careful not to squeeze too much liquid from the udder, which can remove the colostrum."

—Joe Thomas

## Be Prepared

"Regarding foaling, as a first-time midwife, my advice is to sleep in the barn or get a barn cam. Have a foaling kit ready ahead of time with everything you might possibly need, including a camera. Read everything you can about donkey foaling, then read some more."

—Ann Firestone

## Bar the Jack

"When a jenny is lying down foaling, it is not good to have the pasturemate jack present until after the birth. One time this happened to me. The mild jack went crazy when the jenny started to foal. He tried to breed her as she was lying there. I'm sure he was just trying to be helpful but didn't know what to do. He was braying loudly all the time, trying to get my attention that something was wrong. I believe they call this 'crazing.'"

—Pat Scanlan

doctor's appointment, soccer game, or IRS filing deadline pending when you head out to attempt an in-hand breeding. Usually, with in-hand breeding, the jenny is covered every other day during her cycle until she indicates (by kicking, tucking her tail, attempting to run from the jack, or otherwise rejecting him) she has gone "out."

The advantage to in-hand breeding is that the breeder has some control over the safety aspects of the act. The aggressive jack can be muzzled or be otherwise restrained from biting. The recalcitrant jenny can be hobbled or be twitched (a twitch is a pressure device applied to the upper lip of the donkey to release endorphins and produce a calming effect), keeping her from kicking her suitor. The breeder also knows the exact breeding dates, making it much easier to predict and be present for the foaling twelve months later. The main disadvantages are the amount of work and time involved.

## Pasture Breeding

Pasture breeding is a more natural, if a more imprecise, method. In pasture breeding, a jack runs with his harem of jennies and serves them as they cycle. This method usually boasts the highest conception rate because the jack obviously knows best exactly when the jenny is ovulating and needs to be covered. It is also the least labor-intensive approach. It does, however, have its downfalls.

The first is safety. Donkeys can be rough on each other. Jacks tend to do a lot of chasing and biting. If a jenny kicks the jack in a vulnerable spot, she can put him out of the breeding business for months. Jacks can also be extremely aggressive toward the foals of any in-season jennies. In addition, jacks will sometimes choose favorites in a group of jennies, and if more than one jenny

*In this pasture-breeding situation, Jill displays the typical signs of "heat" or "estrus": neck outstretched, ears back, and jaws working in a chewing motion.*

happens to be in season at the same time, he may breed his favorite and ignore the others. It is also nearly impossible to know exactly when a jenny was bred, unless you happen to see it, thereby making it difficult to pinpoint her due date.

## Artificial Insemination

One other method of breeding sometimes used with Mammoth donkeys (or on mares bred to jacks) is artificial insemination (AI). The small size of their reproductive tracts, and the corresponding difficulty in performing the necessary veterinary procedures, makes AI impractical for standard and Miniature donkeys.

If you choose AI, you must closely monitor the jenny for signs of estrus. When the jenny shows heat, a veterinarian usually tracks the progress of her follicle (egg) via ultrasound examination until she is close to ovulation. At this point, if you are buying semen from a jack located some distance away, you notify the jack owner. The owner then collects the jack's semen (the animal has usually been trained to jump a phantom "dummy" or a willing "in-heat" jenny on his end so semen can be collected in an artificial vagina). The semen is extended (mixed with nutrients vital for the survival of the sperm) and shipped out via an overnight carrier. Within twenty-four hours, the "jack in the box" arrives on your doorstep, ready to be inseminated by the veterinarian.

The advantage to using AI is that, in the case of shipped semen, the breeder can make use of a variety of top jacks from all over the country, taking advantage of the very best genetics. The downside is the complexity of timing involved on both ends, the expense of all the veterinary procedures, and the fact that sometimes, for whatever reason, the semen doesn't survive shipping or the jenny just doesn't seem to conceive artificially as well as she does naturally.

Some jack owners do all of their breeding artificially, even at home. One collection of semen can impregnate several mares or jennies at a time, and the risks of sexually transmitted infections are lower than with a live cover. In addition, jacks are extremely quirky creatures, and many have to be coaxed into servicing both jennies and mares. They tend to favor one species and reject the other. By training a jack to collect, either species may be inseminated without the jack's prejudice being a factor.

## Waiting on Baby

Once the breeding process has been accomplished, you then wait to see whether

# Barn Cameras

To the breeders who have discovered their value and are using them, barn cameras are the cat's pajamas. In the past, those breeders would have been padding out to the barn (repeatedly) in their own pjs to check the status of the due-to-foal jenny or been sleeping in them on a bale of straw outside the stall. That's all a thing of the past thanks to the advent of the wireless barn cam. If you have a source of electricity at or near the barn to plug in the camera, a light (a low-watt bulb in a metal "brooder" lamp works great), and a spare television set or computer monitor in the house to plug the receiver into, then you are all but ready to go.

*Jenny Craig poses with her twins, Mason and Dixon. Although extremely common in Mammoth donkeys, twinning usually does not have such a good outcome. It is safer for the jenny to have the vet pinch off one embryo.*

# Foaling Kit

Here are some ideas for items that are handy to have on hand at foaling time.

- Blunt-tip scissors—to break placental membranes if needed
- Bucket—to remove afterbirth
- Bulb syringe, small—to aspirate fluid from the foal's nostrils if needed
- Cellular or cordless phone—to call the vet if needed
- Enema, child size—in case the foal does not pass the meconium
- Flashlight—to provide light if foaling takes place in an area without electricity
- Gloves, obstetric—in case an emergency assist by you is needed
- Oxytocin—must be purchased from your vet and kept refrigerated; administered by shot if the afterbirth is not passed within one to two hours
- Povidone iodine solution—to treat the umbilical stump
- Sterile lubricant—to lubricate your gloved hand
- Thermometer, digital—to take the temperature of the foal or jenny if needed
- Towels, clean—to dry the foal
- Watch—to keep track of the length of foaling

conception has occurred. There is more than one method for determining this; which you use will depend on the breed or size of your donkey and what equipment your veterinarian has. If the vet confirms a pregnancy, your next step is to ensure that the mother-to-be stays healthy so she will carry to term and deliver safely.

## Determining Conception

With Miniatures, determining conception mainly consists of watching the jenny for signs of returning to estrus in approximately three weeks. Few veterinarians are willing to attempt an internal palpation or internal ultrasound to detect pregnancy on the smaller donkeys because of the possibility of rectal tears during examination. A few do offer external ultrasounding. An estrone sulfate blood test can be done anytime past ninety days postbreeding.

With Mammoths, because of a high incidence of twinning, experts strongly recommend that an ultrasound be done fifteen to thirty days after breeding. Twinning is bad news in equines; Mother Nature didn't intend for them to carry more than one foal at a time and will most often abort the fetuses seven or eight months into the pregnancy.

Twins, if they do carry to term, add an extra risk of dystocia (difficult delivery) for the dam, and one or both offspring may be too small and weak to survive. For these reasons, if twins are diagnosed by ultrasound, the vet may suggest pinching off one to give the single remaining embryo—and the mother—the best chance at survival.

Vets who don't own ultrasound machines may diagnose pregnancy by palpation. The vet inserts an arm into the jenny's rectum to feel for a pregnancy through the rectal wall. This method is normally used anytime after about twenty-one days post-breeding; the main downside to palpation is that it will not reveal twins.

Veterinarians who aren't experienced in checking donkeys for pregnancy may have some trouble adjusting to the difference between the internal conformation of a jenny and that of a mare. The jenny's reproductive tract is built on a bit of an up-and-down slope whereas the mare's is more horizontal.

Assuming a pregnancy has been confirmed between fifteen and thirty days, the breeder has approximately twelve months from the last breeding date to await the arrival of the foal.

## Caring for a Pregnant Jenny

The jenny doesn't require a lot in the way of special care during this time. The majority of the foal's development will take place during the last three months of gestation, so don't start upping her feed right away. If you do, you will have a very obese donkey by delivery time, which could make that event harder on her. Just make sure that she receives an adequate, well-balanced diet and all the basic care described in earlier chapters. She should be kept up to date on vaccinations and deworming, although many breeders are leery of administering these very early in pregnancy. Consult your veterinarian for an opinion on this issue. A vaccination most vets will recommend for pregnant

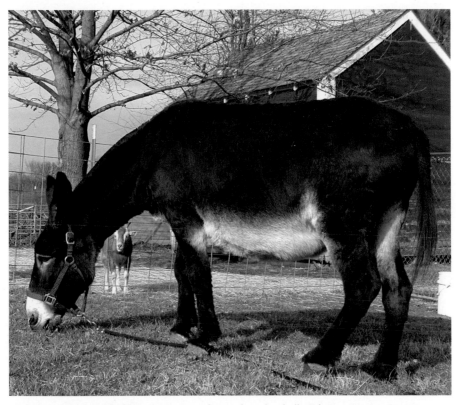

*Heavily in foal, Jennifer has an extreme edema along her belly. Edemas are not uncommon late in pregnancy, especially in older jennies.*

*Pregnant jennies should not be allowed to graze on fescue grass such as this during the last couple months before foaling. Fescue can cause major problems.*

jennies is the one to protect against rhinopneumonitis, a respiratory disease that can cause abortion. The killed variety of the rhino vaccine, which is marketed under the names Pneumobort-K + 1b and Prodigy, is used.

You may continue to ride or drive the jenny until the last few months of gestation. When you begin curtailing her working career, you may finally begin upping her feed to meet nutritional requirements of the swiftly growing foal.

People who live in areas of the country with pastures containing large amounts of fescue grass will want to take the jenny off pasture at least thirty to sixty days prior to foaling. A high percentage of the fescues contain an endophyte (fungus) that can cause a host of problems for pregnant equines, including poor

*A waxy cap has appeared on the engorged teats of a pregnant jenny. This is referred to as "waxing" and, in most cases, will occur within twenty-four hours of foaling.*

milk production; overly long gestations; and thick, tough placental membranes, within which the foal may be trapped and smothered immediately after delivery. Fescue hay should be avoided as well.

Approximately four to six weeks before foaling, most jennies will begin some udder development. Maidens don't usually show as much as the experienced brood jennies do. Around this time, it's a good idea to booster the jenny's vaccines to make sure high antibody levels are passed along to the foal.

## Watching for Foaling Signs

Jennies vary tremendously in the symptoms they exhibit leading up to foaling. Even the same jenny may not do things the same way twice. The following observations and advice are truly general in nature but should provide some guidance. At any point, don't hesitate to consult your vet or fellow donkey breeders for clarification of any troubling circumstances.

The jenny's milk bag will continue to enlarge and become firmer as parturition approaches. The teats will fill out, as well. As the last few days of pregnancy approach, the muscles around the tailhead will loosen and become jiggly. The vulva will loosen and may appear puffy and elongated. You may notice the jenny becoming territorial and grumpy with her herd mates and any squirrels, cats, or birds that invade her space. She may get up and down often as the foal shifts into position for birth.

Within a day or so of foaling, the teats will be fully engorged, with a waxy looking cap on the tips. Don't "milk" the jenny, but express just a few drops several times a day to assess color. When the milk turns from clear and sticky to cloudy or white, foaling is usually only hours away. The jenny will become increasingly restless, often pacing with a distracted appearance. She may paw the ground, get up and down, and urinate and defecate frequently, with a noticeably loose stool. Finally, she will begin to stagger slightly with her tail kinked out to the side. Soon she will lie down and begin to labor in earnest.

## Delivering and Caring for the Foal

At least thirty days or so before the delivery, you should begin planning where the foaling will take place. This should be a clean, safe place away from possible interference by the jenny's herd mates yet easily accessible for you and your veterinarian should assistance be needed. Examine your foaling area with an eye for anything that might prove hazardous for a newborn. Newborn foals have an uncanny knack for finding cracks and crevices to roll under or get stuck in. Many foals have ended up drowning in stock tanks, ponds,

# Saving a Foal's Life

If you harbor any doubts about whether a foal received sufficient colostrum from its dam, ask the veterinarian to draw a blood sample eight to twelve hours after delivery. This is called an IgG test, and it checks to see if the passive transfer of antibodies in the colostrum was successful. If not, your veterinarian may recommend a blood transfusion to ensure that the foal receives immunity from disease.

or creeks, so stalls and foaling paddocks should be free of those. Although the healthiest environment for foaling is a clean, grassy paddock, some protection from the elements may be necessary. Cold and wet conditions can compromise the health of newborn foals, which have poor temperature regulation. A more comfortable setup (especially for those keeping an eye on the jenny) is a spacious pen or stall, well bedded and clean.

## Delivery

The first thing to appear will be a white, fluid-filled bag. Next comes one front foot, closely followed by the other, with the nose and head on top of the legs. If this is not what you see, or if the jenny has been straining hard for twenty minutes or so with nothing showing, call your veterinarian. Until he or she can get there, try to keep the jenny on her feet and walking.

After the legs and nose appear, the jenny may take a bit of a breather while preparing for the effort of expelling the shoulders. When the shoulders are out, the rest of the foal tends to slide on out, although sometimes the hind legs will remain inside until the foal begins to struggle to stand. Once in a while, the jenny gets up and down in the midst of delivery, with the foal hanging out behind her. Although alarming to witness, this behavior is normal.

Donkey births usually take from forty to fifty minutes from the time the jenny begins actively pushing until the foal is born, and most require no assistance from you. If the foal is in proper presentation but the jenny has labored quite some time without progress, you may grasp the ankles of the foal and gently apply downward resistance against the contractions of the jenny. This will be beneficial only during her contractions. The pressure should be

*As delivery begins, the two front feet, one slightly ahead of the other, appear first. The soft, white substance on the end of the hooves is there to protect the dam's birth canal and will fall off when the foal stands for the first time.*

down toward her hocks, not straight out, and should be restricted to traction—no active pulling. Equines cannot stand the kind of calf pulling that cattle can.

When the foal has been delivered, just stand back and observe. Don't cut the umbilical cord; it will tear on its own when the newborn struggles to its feet. If the foal has been delivered with the covering membranes intact, you may pull these away from the head and rub the face and nostrils with a dry towel. Otherwise, leave the jenny and foal to enjoy their bonding time.

Some foals stand up almost immediately; others may take half an hour or so. By then, the jenny should have expelled the afterbirth. It's important that she does so in a timely fashion. If the afterbirth is still hanging from her after a few hours, call your veterinarian. He or she may prescribe a shot of oxytocin to speed the expulsion along. Do not attempt to pull this out

yourself! Doing so can cause lasting damage to the jenny.

## Newborn Care

Spray or dip the newborn foal's navel with gentle iodine, doing so as quickly and unobtrusively as possible, to keep

*A dam and her just-delivered foal sprawl together in the straw. You should not interfere in the process at this point. There is no need to cut the umbilical cord, which will break naturally when the foal stands.*

*The ears of this newborn flop down in a very undonkeylike way. This is not unusual so don't worry; your foal's ears will eventually stand proudly upright!*

from upsetting the new mother; the iodine application will prevent germs and bacteria from entering the bloodstream of the newborn through the umbilicus. Repeat a couple times a day for the first four or five days.

Once you've done this, stand back and watch to make sure the new baby finds the "milk bar." The first milk is rich in colostrum, which has antibodies vital for the foal's health. A foal can only absorb the colostrum during the first twelve hours or so. It is extremely important that the newborn receives and properly processes the colostrum. Without it, the foal can lack immunity to many diseases until such time as its own immune system begins working, usually three or four months down the road.

Watch for signs of the foal passing its first bowel movement, which is a black, tarry-looking stool called the meconium. If you do not witness a bowel movement, be alert for the foal's straining to defecate or keeping its tail held up. It may be necessary to administer a child-size enema to get the "pipes" open and functioning.

First-time donkey breeders are sometimes alarmed when they see their new foal born with wet ears that droop down like the ears of a rabbit. Don't worry! As the ears dry they will stand. In addition, newborn foal legs may be somewhat bent and crooked looking from being crammed into the tight space of the dam's uterus. These, too, will usually straighten on their own during the first few weeks.

# Orphaned or Rejected Foal

Be prepared for any eventuality, including the prospect of having to bottle feed an orphaned or rejected foal. To make sure that such a foal receives the important immunities, collect colostrum from one of your jennies immediately after foaling (within the first twelve hours). You can store the colostrum in a sterile container in the freezer and keep it for up to eighteen months. By doing this, you will have your own supply on hand if sometime in the future you have a rejected or orphaned foal that needs it. The colostrum should be thawed in warm water (never in a microwave) prior to use.

If you do not have colostrum, check with your veterinarian, who may know of other local sources or may recommend a commercially manufactured replacement product. Once you have the colostrum, you need to obtain equine milk replacer. This is usually sold at all farm supply stores; some veterinarians carry it as well. A foal needs to drink 10 to 20 percent of its body weight in milk daily, divided into equal feedings. A bottle with an appropriately sized nipple may be used; however, many breeders recommend bucket feeding as much easier (for both owner and foal).

In general, a newborn should be fed every one or two hours for the first few days. After that, feeding may be spaced every four hours until the foal is six weeks old. Then the foal may be fed every six hours until weaning age.

*An orphaned foal eagerly sucks formula from an "Igloo Mama." To make one, add a rubber livestock nipple to a cooler; the foal can nurse at will, and the formula inside will remain fresh for hours.*

## Raising Baby

For the first few weeks postfoaling, keep an eye on the baby. Make sure it is keeping the jenny's bag nursed down and soft. A hard, swollen bag can be an early indication that the foal is not eating well. In addition, check regularly on the foal's umbilical stump to be certain it is drying up and healing normally. Any bad smell or discharge from that area can be signs of a navel infection that needs immediate treatment. In cool or wet weather, make sure jenny and foal have shelter, especially for the first couple of weeks. Foals are very prone to pneumonia and other respiratory illnesses when chilled by rain or snow.

When foals are nine or ten days old, they frequently have diarrhea. This is tied in with the jenny's "foal heat" (see "Rebreeding" on the next page) and usually passes within a few days. Consult the veterinarian if the diarrhea seems to be combined with any signs of lethargy or lack of appetite.

If you have separated the jenny and foal from other equines, give them one to two weeks by themselves before you reintroduce them to the herd. This allows the pair time to bond as well as time for the baby to gain enough strength to keep up with Mama out in a herd situation. The baby's eyesight (that is, the ability to focus on objects such as fences) also takes some time to develop. The introduction of the pair to a herd should be done cautiously. Often, other jennies and equines will be extremely excited and curious about the newcomer. They may try to chase Mama and baby and cut the baby away from her. This is particularly true of mules, but other jennies are also bad about this (usually young ones that have not had their own foals).

Foals will start nibbling feed within just a few days of birth and will usually continue to eat from Mama's feed pan until weaning age. If the dam is excessively selfish and won't share (and many are), you can build a small pen

*Mother and offspring enjoy separate accommodations from the rest of the herd. For the first week or so, the jenny with the new foal should be kept separate from curious herdmates, who may be inclined to chase the newcomer.*

with an opening just large enough for the foal to enter, with feed troughs set up inside for the baby. This is known as a creep feeder.

Except for the first twelve to twenty-four hours or so, when the important bonding with the jenny is taking place, it's never too early to start handling the new arrival. Baby donkeys are usually born friendly and aren't much trouble to catch or hold on to. More often, at only a few hours old, they are chasing their owners around the stall with their ears back. Following on the heels of their owners with their ears laid back is definitely a donkey thing. They don't seem to be aggressive, but merely making the statement "Hey! You were petting me and now you're leaving, and I don't approve!"

Another unusual donkey behavior is the sinking-back syndrome in foals. Virtually any donkey foal, when touched on the back, will sink downward as if you've hit a very ticklish spot. This disappears with maturity.

Foals should begin halter training when they're a few weeks old. This training is much easier when the foals are small than when they've packed on some pounds and muscle. Haltering the foal and then leading it alongside its dam is usually the easiest way to accomplish the task.

Begin picking up feet and accustoming the donkey foal to having its hooves tapped on and cleaned out. Your farrier will thank you later. Baby's first hoof trim should occur when it's a month or two old. Deworming can start when the foal's about thirty days old and continue on a monthly basis until the foal reaches its first birthday. Vaccinations usually begin when the foal is four to six months old.

Foals may also be weaned around four to six months of age. Do not wean a foal younger than four months except

in the case of a medical necessity such as the dam's "drying up" (losing her milk supply) from accident or illness. Weaning often works best if you keep the foal in familiar surroundings, with familiar herd mates, if at all possible, and move Mama to a new location for a month or so. Preferably, this location will be out of sight and hearing of Junior.

Fortunately, donkeys don't usually kick up nearly the fuss over weaning that horses do. If donkeys have their pals with them to keep them company, they may scarcely be aware of the absence of their dams. Mama may fret for a short time, but if she has companionship where she has been relocated, she too will quickly adapt.

## Rebreeding

Although a jenny will "cycle back" about nine or ten days postfoaling (this is known as the foal heat), she is still in the process of healing and recovering from giving birth. It's preferable to wait for the next cycle (often referred to as the thirty-day heat) to rebreed her, giving her a chance to regroup. In some cases, jennies with foals at their sides will not show heat well, and it may be necessary to wait until the foal is weaned to rebreed.

# chapter EIGHT

# The Hybrid
# Half-Ass

**V**irtually everyone who owns a donkey will eventually have it mistaken for a mule, and vice versa. A surprising number of people (including veterinarians, farriers, and other equine-savvy types) don't really understand that donkeys and mules are two different species.

The mule is a hybrid half-ass created by breeding a horse to a donkey. The donkey has sixty-two chromosomes, the horse has sixty-four, and the mule splits the difference with sixty-three. This odd number of chromosomes means they have no ability to reproduce. Of course, no one has told the mules that they are sterile, and in extremely rare cases, female mules have given birth. Mare mules usually have infrequent, barely noticeable heat cycles of little consequence. The unaltered male mule, while ineffective sexually, is still an aggressive menace and should be castrated at a young age.

The mule, while having longer ears than the horse does, definitely has shorter ones than its donkey parent does. If you compare a donkey and a mule side by side, you will see other physical differences, as well. The mule has a more horsey-looking head and body, and it usually lacks the white points sported by donkeys (the mule's points being tan or brown). The mule has a longer, fuller tail and has a bray that is a combination of the horse's "whee-hee-hee" whinny and the donkey's "hawheeha-whee." The mule's bray is more of a "hawheeheehee." Mules have more-muscular bodies than donkeys do and quite a bit more endurance, strength, and athleticism. Their hybrid vigor makes mules capable of withstanding dryer, hotter conditions than horses can. At the same time, they eat less than horses do (not to mention mules will not gorge themselves into illness as horses would).

The mule's agility and surefootedness make it a virtual equine all-terrain vehicle (ATV). Mules can have you at the top of a steep hill in a heartbeat, with none of the lunging and scrambling of the horse. Mules, which live to ripe old ages, can give

*Randy Cruthis and his mule, Tom, explore trails in South Dakota. Mules are truly equine ATVs, able to travel virtually anywhere.*

long years of service to their owners. Because they inherit the donkey's caution and self-preservation instinct, they seldom require veterinary attention as a result of injury. All in all, it's no wonder mules have served humankind so capably as beasts of burden for millennia.

## History and Background

The history of mules began thousands of years ago, when, either by accident or design, the first mating between horse and ass took place. Historians believe northeast Asia, possibly Turkey, to be the location of the origin of the mule. Mule history in the United States commenced when jacks and mares belonging to Spanish explorers began to populate the Southwest with their hybrid offspring. As discussed in earlier chapters, George Washington, an avid student of agriculture, made great advances in the new industry of mule breeding. His importations of Mammoth jacks from Spain and

## Mule Legalities

Here are a few old-fashioned laws dealing with an old-fashioned hybrid:

- Kentucky law holds it to be contributory negligence for a person to be behind a mule without first speaking to the animal.
- In Lang, Kansas, it is against the law to drive a mule down Main Street during the month of August unless he is wearing a straw hat.
- Mules are protected by law in Ohio, at least to the extent that you can't ride one more than ten miles or to set a fire under one if it balks.

France were the beginning of an agricultural revolution.

For the next 150 years, the mule proved to be as vital to an average farmer's lifestyle as cars are to ours today. Mules began to replace oxen, allowing farmers to farm more land and achieve higher productivity. At the same time, others utilized the mule for freighting and for work on cotton and sugar plantations, in road and railroad building, at logging camps, in mines, and virtually anywhere else that the work was exceptionally strenuous, hot, exhausting, and too rough for horses.

Mules also played a big part in military history, beginning with the Civil War. During the First World War, the U.S. government purchased 571,000 horses and mules for use in Europe; 68,000 were killed in action. Although the military used mules in the next world war, this was the last major conflict in which they were utilized in such large numbers.

The early 1900s were the heyday for the mule in America, with a population of almost 6 million. However, mechanized farming lay right around the corner and brought the mule industry crashing down. The dawn of the tractor proved to be the demise of the mule. Americans forgot the hardy beast that had toiled in the harnesses of their ancestors and helped to build their great nation. By the 1960s, the number of mules had plummeted to fewer than 10,000.

Although their glory days will never be revisited, in the last thirty years mules

*In the early 1900s, miners at the San Lick Mine, near Grafton, West Virginia, pause in their grueling workday to pose with their mining mule. In earlier times, mules were classified by size, with names based their work: cotton mules, draft mules, farm mules, mine mules, sugar mules.*

*This breeding chute owned by Sugar Creek Ranch in Pineville, Missouri, has a front and sides that open quickly in case of a panic situation. It also has a "pit" for the hind feet of the mare or jenny, to make it easier for a smaller jack to reach her.*

have enjoyed a resurgence of popularity. Today, they are being used as recreational and work animals, proving their versatility to a new generation. The U.S. Army adopted the mule as its mascot back in 1899 and still uses it today, and donkeys and mules continue to be used as pack animals for U.S. troops in places such as Afghanistan.

## Making Mules

Folks are frequently surprised to learn that there is a little more to producing mules than meets the eye. Sometimes mules "just happen," when the jack donkey and the mare enjoy a simultaneous "meeting of the minds," but more often, their mating is a carefully planned and orchestrated affair. One added complication to mule pregnancies is neonatal isoerythrolysis, when foal and mare have different blood types.

## The Mechanics

Most jacks do not naturally find mares to be sexually alluring. This unfortunate reality has challenged the ingenuity of mule breeders from the beginning. Breeders have used every trick at their disposal to get the jack to view the mare in a more flattering light. Most breeders agree that the jack should be removed from all others of its own kind at a young age and housed with young horses. The hope is that the jack will begin thinking of himself as a stallion and will act accordingly on reaching sexual maturity.

Matings between jack and mare usually take place "in hand," which means in a controlled environment in which both the jack and the mare are haltered and either held by a lead shank or with the mare tied to a safe surface and the jack led up to serve her. Mares not familiar with longears may be skeptical of the funny-looking suitors and fail to respond.

They are usually led over to where they may be wooed, but not physically molested, by a small pony stallion known as a "teaser." The stallion's job is to sexually excite the mare and get her to signal her readiness to be bred. The mare does so by squatting, raising her tail and holding it off to the side, urinating small, frequent amounts, and "winking" her vulva open and closed.

If the mare shows signs of aggression toward the jack, it may be necessary to restrain her with a lip twitch, hobbles, or a breeding chute. The mare should be unshod on her rear feet, to keep her from injuring the jack should she try to kick. Her tail is usually wrapped, to keep stray hairs from becoming entangled around and cutting the jack's penis.

On occasion, a breeder will have to resort to teasing the jack with a jenny who is in heat to get him excited enough to serve the mare. If all goes well, then the jack will eventually achieve an erection and perform the breeding. Jacks can be excruciatingly slow in becoming aroused, so pack a lunch. Mammoth jacks tend to be take even more time than standard or Miniature jacks do.

If you are purchasing a jack for mule-breeding purposes, ask specific questions about the jack's mare-breeding behavior. This not only helps you know the routine the jack is comfortable performing (and jacks are very much creatures of habit) but also may reveal problems that make a particular jack an undesirable mare-breeding prospect (such as the fact that the owner celebrated two or three birthdays while waiting for the jack to breed his last mare).

Training a jack to have his semen collected for use in artificial insemination has been the answer to a lot of mule breeders' jack headaches. One collection of semen can impregnate several mares, and the collection process need not involve a mare at all. Many vets can assist with collecting and processing semen, and there are university short courses for owners who wish to learn to do the job themselves.

If the jack does his duty and all goes well, the mare will carry her mule offspring for approximately eleven and a half months. This splits the difference between the donkey's twelve-month gestation and the horse's eleven-month one. Care for the pregnant mare just as you would care for the jenny. Keep her vaccinations, deworming, and dental and farrier care all up to date as you await the birth of your bouncing baby mule.

## Neonatal Isoerythrolysis

If a foal's blood type is the same as the mare's, all is well, but if the jack has a different blood type than the mare does and the foal inherits this, a potential for serious problems exists—neonatal isoerythrolysis (NI).

The problem occurs when the incompatible red blood cells from the foal enter the dam's system, usually during birth. The mare will begin making antibodies against the cells, which will be passed to the foal when it nurses the first milk, or colostrum. First foals are rarely affected, but later foals from the breeding of incompatible parents often develop life-threatening hemolytic anemia within their first few days.

If the mare has produced NI in the past, she should have serum tested two to four weeks before her due date to find out if she is producing antibodies to the foal's red blood cells. If so, in order to help save the foal, you must not allow it to nurse from its dam. The foal will need to be muzzled and bottle fed from a mare that has not produced the antibodies and is compatible with the foal. It is preferable that the colostrum be taken from a mare that has never had a mule foal. Fortunately, the level of antibodies will gradually decrease over twenty-four to thirty-six hours, and the foal will not absorb the antibodies after that time and may then be allowed to nurse from its mother.

If the mare has not been checked and produces a mule foal that is weak, jaundiced, and lethargic in its first few days, suspect NI. A blood transfusion may save the life of a foal, but it is not always successful.

## Mule Behavior and Handling

If there is one word that describes mules, it's *smart*. Despite history's long characterization of the mule as stupid, ornery, and bull-headed, cognitive tests place the mule's intelligence higher than that of either of its parents. Mules are bright, curious, durable, strong, self-preserving, sensitive critters. A good mule inherits the best of the donkey's characteristics—intelligence, caution, and hardiness—with the best of the horse's—strength and energy—to blend into an ideal using animal.

Because mules are so smart, it is advisable to start working with them from a very young age. They benefit immensely from imprinting at birth and from regular human companionship and handling from foalhood on. It's often said that mules don't "grow a brain until they're ten," and it's true that they seem to mature mentally at a slower rate than horses do.

Mules tend to be slow to trust humans, but once you earn that trust, you will have loyal and steadfast partners. Because mules mature slowly, it's best to start saddle and harness training them a little later than you would a horse, and their lessons need to be custom designed with

# What About Hinnies

A mule is created by the mating of a jack donkey with a mare horse. The flip side of this, a stallion bred to a jennet, produces a hybrid that is basically visually indistinguishable from a mule, a hinny. Some people believe that hinnies are a bit more horsey looking than mules, but this is very debatable. Due to chromosomal glitches, it is more difficult for a jennet to successfully get in foal to a stallion and carry the fetus to term. Hinnies are a true rarity, and most folks who attempt to set up a hinny-breeding program become discouraged and abandon the project.

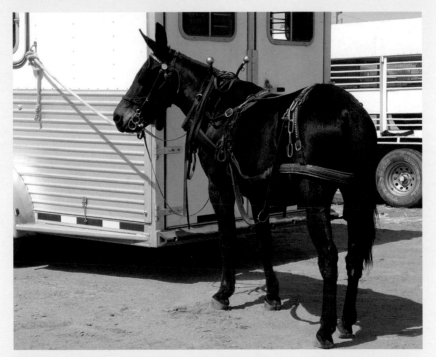

*This hinny, TNT's Dyna-mite, is registered with the ADMS as a mare hinny, registered with the American Part-Blooded Horse Registry as a half-Morgan, and registered with the American Warmblood Society. Her breeder was Barbara Nees of TNT Mules & Morgans.*

*Tom Cochran and his mule Coal practice working cattle. Mules excel in the same disciplines that horses do.*

the peculiarities of their species in mind. They're quick learners, bored by endless repetition, and they adhere very rigidly to routine. It's all about the routine when it comes to mules.

Mules, never looking in a mirror and discovering otherwise, go through life believing they are horses. Their mothers were horses, and thus they have no reason to doubt this is true. They usually have the horse's snooty, superior attitude toward donkeys but are crestfallen that horses never seem to accept them as equals. They are extremely attached to horses, in this never-ending quest for acceptance.

This is why many mule outfitters can turn their pack strings out to graze freely at night in the presence of a "bell mare." The mules will not leave the mare, which has a bell tied to her to make her easy to locate and catch the next morning.

Mules have a reputation for being more "cold-backed" than horses. This means they are fussier about being saddled and cinched up and more prone to bucking when first mounted. Their reputation for stubbornness and balkiness, like that of the donkey, is actually a misinterpretation of the inborn cautious nature they possess. The mule will just flat out refuse to do anything that it feels is not in its best interests. The problem occurs when the opinion of the handler and the opinion of the mule differ. But if

# Coon Jumping

Coon jumping is an entertaining event unique to donkey and mule shows. The idea originated from coon hunters, who often rode mules at night in pursuing their quarry. When they came to a fence, they would dismount, throw a jacket or saddle blanket over the top of the fence to make it more visible, and jump their mule over it from a standstill. Mules (and donkeys too, although perhaps to a lesser degree) are capable of tremendous jumps from a flat-footed start. Horses are not physically capable of this and should not be asked to attempt it.

Many donkey and mule shows have incorporated coon-jumping classes into their offerings. Each species usually competes against its own kind, but classes are not subdivided by the animal's size. In many parts of the country, the animals compete with nothing but a halter and a lead rope. However, in the southwestern states, they frequently sport a saddle, as well. A freestanding jump is required, from a chalk-marked box on one side, up and over a sliding vertical rail or crossbar supported by pins or pegs, which will fall upon contact. The handler may stand on either side of the jump but must hold on to the lead-rope at all times. Amazing feats of athleticism have been accomplished in this event. Some years back, a 54-inch mule from Texas coon-jumped 78 inches (6½ feet) in competition.

*Janis Flynn and large standard donkey Sweet Betsy compete in a coon-jumping class at a Midstates Donkey and Mule Club show. Coon-jumping classes are unique to donkey and mule shows; horses do not have the same ability to jump from a flat-footed start.*

# Mules, the Hybrids

Here, our experts talk about understanding and dealing with that unique member of the equine family the mule.

### Think About It

"When we were new mule owners, the thing that surprised us was the way mules think and process things. You certainly can't 'make' them do what they don't think is a good idea! This made us think about the way we introduced things. Positive vs. punishment—reward, reward, reward the good."

—Terry Lupien

### Keep a Sense of Humor

"The main thought that came to my mind about raising mules is to do it with a sense of humor. These hybrids bring out the best in the owner and sometimes the worst. They will make you laugh most of the days and will sometimes send you to the porch with tears streaming down your face.

"They respond to all loving touches and firm (but needed) reprimands. All living creatures need this type of guidance. I taught first grade for thirty years in the public school system. Many people have asked me how I have so much patience. My answer was simple, 'I work with mules.'"

—Peggy Reed

### Be Kind

"Mules have taught me so much since [I've had] the rescue [Save Your Ass] and [dealt] with a lot of them. They are so smart and so very, very sensitive. The latent learning is amazing. Give them a lesson one day, skip a day, and go back the day after, and it seems like they almost always have thought about it and get it.

"Be kind all the time. That should go for all animals, of course, but the mules especially as they are so sensitive and smart. That can work against one if kindness is not employed always."

—Ann Firestone

### Start Out with a Seasoned Mule

"We've had mules since the '80s. We started out with an aged pair of draft Belgians that were already trained to pull a wagon and plow. Anyone new to mules would be wise to start out with seasoned mules. Most folks say that's around seven years of age for the mule. A mule must decide it trusts you as an owner/rider. Building that trust is the most important training that mule will receive. A relaxed mule that is confident in its rider makes a life-long steed."

—Lynn McMillan

## Be Aware

"I think that people need to be aware that a mule is not a long-eared horse. Not everyone who enjoys working with a horse is going to be happy with a mule. I know of several people who have admired mules. Instead of buying a well-broke mule, they breed a mare to a jack and raise their own. Quite often, they are never able to do anything with the mule. There is a reason why well-broke mules are so expensive."

—Janis Flynn

## Don't Get in a Hurry

"You have to remember that a mule is also half horse and has the reactive side of the horse and not the donkey, but also will stand and fight more like a donkey than a horse if it feels cornered or threatened. With mules, you have to earn their trust; it is very seldom given freely and never from a previously unhandled or abused mule. However, you will find that most mules will respond and work well for food. Treats, lots of treats, a calm, quiet presence around them, be fair but firm with your rules, and lots of love.

"Mules mature physically and mentally a lot slower than a horse does. Even at three or four years old, a mule is still a baby mentally. In my opinion, many mules are ruined by pushing them too fast too early."

—Vickie Rauh

the mule indicates something shouldn't be done, it's a good idea for the handler to reassess the situation before arguing.

In addition, there's the mules' famous reputation (shared with donkeys) for kicking. It's true that mules are extraordinarily dexterous and athletic and can seemingly kick in any direction and at lightning speed. However, a gentle, well-handled, and well-trained mule or donkey seldom kicks. Often the root of this problem goes back to the fact that mules are suspicious of strangers. They don't appreciate being handled or having their feet picked up by someone they haven't been properly introduced to and personally approved.

Traditionally, female (referred to as molly or mare) mules have been the preferred sex for owners. They are reputed to be easier to get along with and less devotedly attached to horses than the male (john or horse) mules.

Mules of both sexes often have an obsessive, fanatical interest in foals of all kinds. When new foals arrive, make sure to keep them separated from mules until you've had a chance to adequately assess the mules' intentions. John mules have a tendency to want to attack foals, while mollies are more likely to attempt to steal and "mother" them.

## Types and Uses of Mules

Mules come in a variety of sizes, shapes, and colors. Depending on the size, color, and breed of horse and jack used, mules can range from 36-inch dun miniature mules to 18-hand sorrel drafts, and everything in between. Whatever your fancy, there is a mule for you.

## Miniature Mules

These pintsize longears pack a lot of power in a small package. Produced by breeding miniature jacks to miniature

*A mini-mule "three-up" participates in the annual Mule Day parade in Columbia, Tennessee. Mini mules are extremely strong for their size.*

horse mares or small ponies, miniature mules are popular for use as pets and as driving and pack animals and in petting zoos and circuses. They come in a range of colors. Miniature mules can be registered with the American Donkey and Mule Society or with the organization devoted strictly to them, American Miniature Mule Society. The latter offers registration for class A mules (under 38 inches) or for class B mules (over 38 inches but under 48 inches).

## Saddle Mules

A saddle mule can be of any breed or type. Most, however, are the product of a jack crossed on the more popular breeds of horses, such as Quarter Horses,

## Mule Celebrations

Small town festivals and celebrations honoring the mule are held all over the country. Three of the biggest, best-known, don't-miss events are Bishop Mule Days in Bishop, California; Columbia Mule Day in Columbia, Tennessee; and the Great Celebration Mule and Donkey Show in Shelbyville, Tennessee. If you ever get a chance to attend one of these, you will become immersed in all things mule and donkey. Entertaining, informative, and just a whole lot of fun, they will appeal to any longear enthusiast.

*These New Mexico mules can carry their riders all day. Riders prize mules as saddle animals for their good sense, surefootedness, and endurance.*

Paints, Thoroughbreds, and Appaloosas. Those seeking a gaited mule will look for crosses from Tennessee Walking horses, Missouri Foxtrotters, and Paso Finos. Some breeders concentrate on certain colors and want mules from Appaloosas or Paint mares to pass on the color patterns of those breeds. Today's modern saddle mule is light-years away from the jug-headed, knock-kneed, ornery cuss of yesteryear.

In recent times, extremely selective mule breeding has raised the industry to a whole new level. The combination of high-quality mares and jacks are producing mules that excel at virtually every equine discipline: English and Western riding, dressage, pleasure driving, roping, team penning, trail riding, racing—if you can name it, a mule can certainly do it.

## Draft Mules

These are the gentle giants of muledom, produced from Mammoth jacks and draft horse breeds of mares (usually Belgians and Percherons). The Amish use work teams on their farms, and driving enthusiasts enjoy draft mules for use in recreational wagon trains and carriage driving, and for competitive pulling. Belgian mules are nearly all sorrel in color, while Percheron mules are gray or black.

Mules can be registered with the American Donkey and Mule Society, the American Mule Association, the North American Saddle Mule Association, or for spotted stock, the American Council of Spotted Asses.

## Finding Your Mule

The search for a good mule needs to be undertaken with a great deal of research,

*Breeders produce big sorrel draft mules, such as this pair, with Belgian mares and Mammoth jacks. Draft mules are popular for recreational driving, pulling competitions, logging, and farm work.*

forethought, honest appraisal of your plans and abilities, and in most cases, a considerable amount of money. It may come as a shock to many horse people migrating over to the world of longears, but good mules don't come cheap. An incredible amount of time and work goes into producing sensible, well-trained mules with good conformation, and naturally those who have put forth that effort expect to profit from it.

As a buyer, you must bear in mind that a good mule is very, very good. It will serve you tirelessly and faithfully in whatever way you ask. A bad mule, by contrast, will get you hurt. A mule that, whether through poor training or harsh handling, is not a team player, is quick and smart and strong, and won't hesitate to demonstrate all three characteristics as it ejects you from the saddle or runs

off with your cart. A penny saved on the purchase price will not seem like such a bargain to you if you have to spend it (and quite a few more) on doctor bills down the road.

If you're not familiar with mules, try to find someone who can mentor you and help you shop. Local classifieds, Internet classifieds and breeder listings, industry periodicals, and word of mouth are all good resources to help find your perfect mule. Auctions are usually not. With mules especially, it's best to spend some time with your prospect and, if at all possible, try it out for its intended purpose. Take a test ride or drive. Remember that you're new to Mr. Mule, and he's probably not going to take to you immediately or perform flawlessly, but you should get some idea as to whether he suits you.

Most saddles are made for a horse's conformation. If you're buying a saddle mule, look for good withers (shoulders). This area at the base of the mane where the neck meets the shoulders needs to be slightly prominent. Many mules inherit the flat, broad withers of their donkey sires, and this build makes saddle fit an aggravating problem. If the animal has no withers to speak of, most saddles will slide forward, causing pinching and soreness. In addition, the girth holding the saddle in place will work itself up behind the front legs, in the "arm-pit" area of the mule, and cause nasty, open sores referred to as galls.

Fortunately, due to increasing popularity of longears as riding animals, there are a few manufacturers offering mule saddles. Most folks riding mules and donkeys also equip the saddle with a crupper or britchen to help keep it from sliding forward. The crupper is a loop that fits around the base of the tail and fastens by a strap to a ring on the back of the saddle. The britchen is a more involved, harness-looking affair that fits around the hindquarters of the animal. Cruppers are usually adequate except in the steepest, most rugged terrain; there a britchen is recommended.

Whatever type of mule you're looking for, try to find one with a gentle, kind eye. Mules reveal a lot about their nature through their facial expressions. A mule that is fearful and distrustful will show it in its eyes and tense, stiff posture. A person new to mules is always better off starting with an older, friendly, well-trained mule than with a young "project" or a mule with issues. Issues can be overcome with time, work, and patience, but the "been there, done that" mule makes the best and safest teacher.

*Prize-winning draft mule colts check out visitors at the Terry Givens farm in Missouri. Givens finds a ready market for these colts with Amish teamsters.*

Hi my name is Daisy. I am a Mammoth Jenny Donkey. I live at the Peterson Farm in Mansfield. My job is to protect the cattle herd from wolves, coyotes, etc...

# chapter NINE

# Making Money with Donkeys and Mules

**D**onkey breeders don't generally retire to a tropical paradise from the millions made in the industry. While other types of livestock can be raised for various commercial commodities (meat, wool, milk, eggs), the donkey (in America anyway) cannot. There is one potentially profitable business venture a donkey owner might undertake: breeding high-quality speciality donkeys. Yet not everyone can prosper in the breed business, which takes a great deal of planning and work and some serious capital outlay. For most donkey owners, a breeding business is impractical. The good news is that there are other ways donkeys can at least contribute to their upkeep, if not put their owners in the black.

## Raising and Selling Your Donkeys

Raising and selling top-of-the-line registered Miniature or Mammoth donkeys can be one of the most financially rewarding enterprises related to longears. To be successful in this arena, you must first learn as much as you can about your product and then about your market. Then you will need to develop clear-cut goals for a breeding program that will set you apart from your fellow breeders. Think about certain characteristics that you could strive to excel in—such as conformation, size, disposition, color, or bloodline. While you should never sacrifice everything else to acquire a particular characteristic, to succeed as a breeder you do need to carve out a specialty niche for yourself in the marketplace. Once you've settled on your goals, go looking for the right donkeys. It is crucial that you buy the best animals your bank account will allow to serve as the foundation for your program.

Once you've developed the product (your special donkeys), the next step is to market it. Figure out who your buyers are and advertise through the most appropriate media. Today, Web sites have become a must-have in advertising; they offer

*Donkeys who have less common coloring, such as this spotted Mammoth jack colt, always find a good market. Look for a niche that needs filling when considering a breeding program.*

worldwide exposure and the best value for your dollar. A Web site can create a strong first impression of your business, so hire a professional to set up your site if you don't have the skills to do so. Other places to advertise include trade magazines and local newspapers and at local tack and feed stores.

In addition to advertising, get your donkeys noticed and admired. Haul them to public functions such as parades, seasonal events and festivals, shows, and even visits to elderly residents in nursing homes (not only will your donkeys be seen but they'll also brighten some people's day). Have business cards made, and hand them out freely. Someone may contact you years later or pass your card along to a friend who will be interested.

Keep supply and demand in mind. When the market is flooded or stagnant, lower the asking price for your donkeys or curtail their breeding until demand rises. Another way to control an overcrowded market is to geld surplus jacks and make them available as pets. In any case, few jacks are of high-enough quality to keep as breeding animals. You don't want mediocre donkeys!

Most of all, be an honest and ethical breeder. Do not put profit before the welfare of your donkeys.

## Other Income Ideas

If you have no interest in breeding top-quality donkeys, you can pursue other ways to bring in a bit of much-needed income from your animals. You won't get rich from any of these endeavors, but you could cover some of your more outstanding donkey-owning costs. Although you can't put a price on the enrichment

*A striking Miniature donkey and his owner compete at the Houston Livestock Show and Rodeo. Minis at the highest level of competition bring top dollar, even in the depressed equine market.*

donkeys bring to your life, you can find creative ways to help defray expenses.

## Raising Mules

There is a expanding market for mules. Over the past twenty years, saddle mules, in particular, have become exceptionally popular and have steadily improved in quality. Trail riders, packers and outfitters, and show exhibitors provide a steady market for saddle-type mules. Although fewer in number than saddle mules, draft mules are also in demand today, for pulling competitions as well as for recreational driving.

As discussed in chapter eight, mules are produced by breeding a jack to a mare, and hinnies by breeding a stallion to a jennet. Because jennies have more trouble conceiving and carrying a hinny to term, hinnies are rarer than mules. Given this

problem and the fact that mules and hinnies look and act virtually the same, breeding for mules makes the most sense.

To sell mules, follow the same methods used for selling donkeys: advertise in trade publications, on equine Web sites and other online sales sites, at tack and feed stores, and through words of mouth with other mule folk.

### DID YOU **KNOW**

In Europe, there is a good demand for donkey milk. Although it is used mostly in soaps and cosmetics, people also drink it. Donkey milk hasn't really caught on yet in America, but perhaps donkey dairies will become the next big thing!

## Offering Stud Service

If you have a jack of exceptional quality—preferably a registered one with a good pedigree—you may realize some income by offering his breeding services to mares or jennies. The jack should have some sort of show or working career or other public exposure to let the world know of his existence and sterling qualities.

You will have to lay out money for equipment, setup, advertising, and shipping. While your jack waits back in the boudoir for the dates to come rolling in, you will need to build facilities for prospective brides (for example, stalls, pens, breeding chutes) or a setup for collecting semen and preparing it for shipment.

It's a good idea to draw up a breeding contract, which will signed by both jack owner and customer, to set forth conditions regarding what is expected.

When you are ready to open your door to customers, start placing ads in trade publications on the Internet, telling everyone why they should use your incredible jack for breeding.

## Offering Donkey Training

You could hang out your shingle as a donkey trainer. As has been discussed throughout this book, donkeys are not "long-eared horses," and horse trainers not only are often reluctant to take on donkeys but also may not be successful

*Audrey and her rider, Anita, step out on the trail at Wrangler Campground, Land Between the Lakes, Kentucky. Mules are in demand now for riding, packing, and other such activities.*

*A mature breeding jack such as this one can generate income for his owner through stud service to mares and jennies. The jack must be of exceptional quality.*

*Pauline and her rider learn team penning at a clinic. Well-trained Mammoth donkeys, such as Pauline, are in great demand and short supply.*

*In the pasture, a standard donkey guards cattle. Ranchers and farmers buy donkeys to protect newborn cows, lambs, and other livestock from predators. Jennies and geldings are best.*

with them. A person with an understanding and appreciation of the species can get the most out of a donkey.

The demand for Mammoth saddle donkeys has been growing by leaps and bounds in recent years, with a corresponding demand for people who are qualified to train them. If you have the ability to train, you will likely have no shortage of customers.

Miniature donkey owners are very fond of driving to cart, so the offer of driving training should also be well received by the public.

## Selling Pack Animals, Pets, and Livestock Guardians

You can generate income selling donkeys to farmers and ranchers as livestock guardian and pack animals as well as to other buyers as pets. Modestly priced standard donkeys are used for the practical functions. Many people will bypass you and get their standards by adopting BLM burros. Others, however, prefer to buy a domestic donkey that has been handled, lead broke, trained to have its feet trimmed, and so on, when looking for a pack animal or livestock guard.

Advertising in feed stores, sale barns, co-ops, and wherever livestock owners congregate will help get the word out that you supply donkeys for their purposes.

# Selling Outside the Box

There are some less-conventional methods to make money with donkeys. One steady by-product of donkeys is their manure, which might be marketed by the enterprising owner to gardeners and nursery owners as organic fertilizer. An owner could also keep a rental string of Mammoth donkeys and offer the extremely unique "all-donkey trail ride." Donkeys could be leased to individuals for use in nativity scenes during the winter or for brush-clearing service in the summer.

*Jassper tells Santa that he has been a good boy and wants a bow-tied bale of alfalfa for Christmas. Donkeys are frequently wanted for nativity scenes and seasonal festivals.*

*A set of stocks, such as these at Bess Jackstock Farm, can be useful for restraining donkeys during veterinary, farrier, and grooming procedures.*

## How to Save Your Ass—Financially

Having discovered that the best way to make a million dollars with donkeys is to start with two million, you may prefer to look at ways to be more economical in your donkey husbandry.

## Doing Farrier Work Yourself

A donkey owner can save money (especially with multiple donkeys) by doing some of his or her own farrier work. Although few owners will be confident enough to do their own shoeing, many can learn to do trims or at least to use a rasp to file hooves and lessen the need for frequent trims.

Ask a farrier, a vet, or an experienced layperson to show you some techniques or study one of the many books or videos on the subject. At least once a year, have a farrier do a trim to examine the job you've been doing, correct any flaws in procedure, and give you tips.

## Performing Veterinary Procedures

The average owner can learn to perform certain veterinary procedures. Under the tutelage of a vet, you can learn to administer routine vaccinations, antibiotics, and dewormers.

Learning to recognize when veterinary assistance is needed is one of the

*Big round bales of hay rest under a protective roof. Growing your own hay can save you money; you may be able to get a local farmer to harvest the hay for a share of it.*

most important aspects of donkey ownership, from the standpoint of potential financial savings as well as the well-being of your stock. Learn the red flags that mean "Call the vet ASAP!" and the yellow flags that mean "Wait and see."

If at all possible, haul your donkeys into the vet's office for treatment to avoid costly farm-visit charges, or see if your vet will divide travel costs between you and several neighbors who coordinate to use his or her services at the same time.

## Growing Your Feed and Buying Wisely

Growing your own feed can take some sting out of the cost of donkey keeping.

Hay, especially, can be fairly easily grown and produced. If you don't want to invest in haying equipment, a local farmer may be willing to bale for you on shares (he may keep one third or one half of the hay in exchange for his services).

Be a savvy shopper when buying tack and equipment. It doesn't have to be new, just well cared for. Web sites such as eBay and Craigslist can be potential goldmines when it comes to good used saddles, bridles, harnesses, carts, show clothes, and many other items that are exceptionally expensive when purchased new. Watch ads for tack swaps and your local trading post–type papers for used water tanks, hay rings, fencing, and farm materials.

# Making and Saving Money

Our experts weigh in on how you might make some money on your donkeys and mules and how you might save some.

## The Best

"Probably the best advice I was given when starting to buy my first donkeys was 'Buy the best animals you can afford. It does not cost any extra to feed a good one.' This old adage carries a lot of truth. Feed, wormer, farrier visits, vet calls, and most of the other expenses are the same, regardless of the animal. However, the returns generally will be higher with a better-quality donkey. Offspring of quality parents will usually sell quicker and for more money. Good-quality animals also will outperform poorer-quality ones in the show ring."

—Joe Thomas

## An Ounce of Prevention

"Prevention is the best way to operate with donkeys. Keeping them healthy is better on the pocketbook."

—Lynn McMillan

## Street Feeding to Detachment

"Have 'street-feeding days'—put all the critters out on a line so they can eat the fence-row grass and save the pasture. Use smaller water tubs with automatic fillers—keeps the pump from running all the time. Irrigate the pastures at night—less evaporation, and we use a pump to water our pastures. Learn to trim your own feet—well, not your own, but your herd's feet—no farrier bills. Cut down the size of the herd so you can have some hay to sell instead of using it. Don't get so blamed attached that you won't sell one when you can!"

—Robert Auge

## Added Value

"My advice would be, either in trying to sell or save money, any added value is important. Example: extra training like driving, riding, showing will help in sales. Well-groomed, good pictures will go a long way. Poor pictures will hurt Internet sales. Well-written ads and a Web site are helpful."

—Marla Ellestad

## Quality Breeding

"To make money in jackstock today, the most important principle is quality breeds quality. Whether you are raising jackstock for riding, harness, or show or to increase your breeding stock, breed the best jennets to a jack that complements them in conformation, color, and attitude. In this day and age, everyone wants jackstock with these qualities. Don't raise donks just to raise donks."

—Randy England

## Stockpiling, Sharing, and Mail Order

"We stockpile our manure in the dry lots both spring and fall and use the old dirt for the garden. In winter we

share water tanks between fences, using one heater instead of two. We are going solar on the large turnout rather than using electric. We are buying our oats from a local farmer at about ten bucks per fifty pounds. Vaccinating using tanks of ten doses is cheaper by mail order than through the vet. Buying dewormer by mail order is cheaper than at the local feed store."

—Sue Elliot

### Homemade Repellant

"This summer, instead of buying expensive fly sprays for the donkeys (about twenty dollars a bottle and not lasting the fourteen days as advertised), I elected to mix ½ cup of raw apple cider vinegar (1 gallon under seven dollars) as part of the water used in their daily beet pulp recipe. It has kept the majority of the flies off them, especially on the belly, and I don't believe they are being stung by bees or bitten by the flies. It has worked very well. I also use Tea Tree Oil mixed in a little Vaseline as a wound treatment, but keep it away from where [the donkeys] might rub it into their eyes."

—Kristi Kingma

### Gross (but Reusable) Traps

"Use reusable fly traps. They are so gross, but you need only to put a piece of meat in a trap as an attractant and the flies will love it. Dump it as needed, and just fill halfway with water and a piece of meat."

—Ann Firestone

# Acknowledgments

This book would never have materialized without the faith and the encouragement of several people, including BowTie Press editors Andrew DePrisco and Jarelle Stein; as well as Sue Weaver; Gerry Griffiths; and various friends and family who believed that a middle-age housewife/donkey breeder with no writing background could become an author.

I have to thank those individuals who have generously shared their knowledge and taught me so much about donkeys, especially Sandy Connelly and Bill Gunter of Roswell, New Mexico; Mary Hubbard of Rowley, Massachusetts; Joe Thomas of Readyville, Tennessee; Dr. Tex Taylor of Bryan, Texas; Randy England of Charlotte, Tennessee; Pat Scanlan of New Braunfels, Texas; and Jim "Throcky the Answer Man" Ensten of Ohio, just to name a few.

Huge, huge thanks goes to all of those who responded to my requests for "Advice from the Farm," including many from my "family" at the Yahoo Internet discussion list "Mammoth Donkeys." They are a never-ending source of friendship, information, and support.

Thanks to my sidekick and friend, Bobbi Ward, for accompanying me on countless trips to visit farms, attend sales, show donkeys, and take pictures and for being my general "right-hand woman."

Thanks to the best veterinarian that a donkey breeder could have, Dr. Harold Bristow, who has never failed to be there when I needed him, and to my fantastic farrier, Danny Merriman.

Last but certainly not least, thanks to my husband, Roger, and my son, Noah. They gave me much-needed advice and feedback as the book was being written, helped stage photographs, and put up with a frazzled and distracted wife/mother during the months I worked on the book.

Finally, I must thank the donkeys themselves, for enriching my life in ways too numerous to mention. Happy brays!

# Appendix:
# Afflictions at a Glance

There are certain diseases and abnormalities you should watch for in donkeys, even one basically healthy. Here are some of them.

## Choke

Choke occurs in donkeys when partially chewed food gets wadded up and lodged in the esophagus. It can be caused by the donkey's not chewing food completely, the type of feed (sometimes pelleted feeds are more prone to compacting), dental problems, or a partial obstruction of the esophagus by a tumor or old injury.

A donkey showing choke symptoms is an alarming sight. Typically, the donkey's neck will be outstretched, with its eyes fearful and confused. Saliva will be running from its mouth and nose. The obstruction may or may not be visible externally. Although unable to swallow, the donkey can still breathe so is not technically choking. Most of the time, the situation will resolve itself within a few minutes with no intervention. You can attempt to massage the throat and neck area in an effort to break up the lodged mass. Squirting a needleless syringe filled with water into the mouth may help wash the mass down, as well. If these efforts fail, call the veterinarian. He or she will sedate the donkey and use a stomach tube to break up the obstruction.

To prevent recurrence of choke, try to identify the causative factor. In a case in which the donkey is eating rapidly and bolting its feed, put some large rocks in the feed pan to slow down consumption. Make sure any dental issues are addressed. If you are feeding pellets, switch to grain; some donkeys seem to handle grain better. Always make sure the donkey has access to plenty of water.

## Colic

The term *colic* refers to the symptom of abdominal pain rather than a specific illness. Signs of colic can include pawing, stretching, turning the head back to look toward the flank, refusing to eat, repeated rolling, little or no fecal output, and depression. Donkeys tend to be much more stoic in colic cases than horses do, so you will need to observe your donkeys closely. If they display any of these symptoms, don't hesitate to consult your veterinarian.

Colic can have many causes. Often it is some sort of obstruction, which may be formed by impacted food or a kill off of a large number of parasites following a deworming. The donkey's ingestion of a large amount of sand by grazing or being fed on sandy soil can also lead to an impaction. Gas buildup, enteroliths (intestinal stones), and a twisted or telescoped section of intestine can also be causes of colic.

Incidence of colic can be reduced by providing plenty of fresh, clean water at all times; restricting access to rich feed; making changes in feed gradually; not feeding directly off the ground, particularly on sandy soil; and providing regular deworming and dental care. Feeding at the same time every day can be beneficial, as well.

## Cushing's Disease

Cushing's disease is a condition involving the pituitary gland. The pituitary gland becomes overactive and causes the adrenal glands to overproduce cortisol. Excessive cortisol causes a plethora of health problems.

Symptoms include long-hair coats that are kept year-round, lethargy, sweating, excessive drinking and urinating, and fertility problems. There is also an increased risk of laminitis, and the elevated cortisol levels interfere with the immune system, making infections, such as hoof abscesses and skin conditions, more common.

Cushing's is diagnosed with blood tests. No cure exists, but most cases can be managed with drug therapy and diet. Reducing carbohydrates can be beneficial. There's no evidence that Cushing's can be prevented.

## Equine Influenza

Just as in humans, influenza in equines is caused by a virus. The virus is spread via airborne particles, which can be carried on hands, shoes, feeding equipment, and other everyday items. The virus spreads quickly and is quite contagious.

Symptoms include fever, cough, nasal discharge, and loss of appetite. Young animals are usually the most severely affected, and in an unvaccinated herd, virtually all animals will be infected to some degree when exposed to the virus.

Most equines make a full recovery from influenza. The main threat of complication is secondary infections, such as pneumonia, which can be quite serious in donkeys. Vaccination is the best form of prevention, although influenza vaccines are short lived and booster shots need to be administered more often than with some other vaccines. When dealing with a herd outbreak, maintain good hygiene to lessen the effects of influenza.

## Equine Protozoal Myeloencephalitis (EPM)

EPM is caused by the donkey's coming in contact with food or water contaminated by a protozoal parasite known as *Sarcocystis neurona*. The protozoa are carried in the feces of their host, the opossum, which is why EPM is sometimes referred to as the possum disease. The protozoa attack the donkey's central nervous system, and if left untreated, they can cause damaging long-term effects.

Symptoms of EPM may vary, depending on the severity and location of the lesions that form in the brain, brain stem, or spinal cord. They can include: ataxia (incoordination); spastic, stiff, stilted movements; muscle atrophy; paralysis of the muscles of the eyes, face, or mouth; difficulty swallowing; and tilting of the head combined with poor balance.

If the vet suspects EPM, he or she will do a spinal tap and run a CSF (cerebrospinal fluid) analysis. This test can be useful in that it will indicate that the donkey has been exposed to the parasite, but testing positive does not necessarily mean that the donkey has or will develop the clinical disease. If clinical symptoms are present, the vet may prescribe one of several drug regimens. Overall, with treatment about 25 percent of diagnosed equines respond well enough to return to their original functions.

There is little that can be done to actually prevent EPM. Keeping your feed room and containers cleaned up and sealed to discourage rodent activity is a top priority. There is some evidence that the protozoa do not survive the heat from processing cereal grains and extruded feeds, which may make them a better choice for your feeding program. Discuss vaccination options with your veterinarian.

## Equine Encephalomyelitis

The three most common varieties afflicting equines in the United States are the eastern (EEE), western (WEE), and Venezuelan (VEE) versions of this viral disease of the brain. Like West Nile virus, encephalomyelitis is a mosquito-borne illness that attacks the brain and spinal cord.

The clinical symptoms of EEE, WEE, and VEE are very similar: fever, depression, incoordination, loss of appetite, facial paralysis, and seizures are all common. Diarrhea may be present in VEE. The main differences between the three varieties are in the death rates. Mortality ranges from 75–90 percent for EEE, to 19–50 percent for WEE, to 40–90 percent for VEE.

A positive diagnosis is usually made when an animal displays clinical symptoms during mosquito season in a region where these diseases are prevalent. Supportive therapy such as fluids, anti-inflammatories, and steroids may aid recovery, but survivors are often affected with permanent brain damage.

Good vaccinations to protect against encephalomyelitis are available. They have low reaction rates and are usually effective for from six months to a year. Vaccinating just prior to mosquito season will offer your donkeys protection during the most dangerous part of the year.

Mosquito control is also advised. Dump stagnant water from tanks, buckets, and barrels around the farm. Try not to house any animals in low-lying, swampy areas. Apply fly repellents (which also work to repel mosquitoes) to the donkeys, particularly at dusk when the mosquitoes are most active.

## Equine Infectious Anemia (EIA)

EIA is also known as swamp fever. It is a retrovirus transmitted by flies, mosquitoes, and other bloodsucking insects. There are three forms of the virus:

Acute—the virus is active and doing damage to the animal's immune and organ systems.
Chronic—the animal may swing back and forth between an active disease state and remission.
Inapparent—the animal carries the virus but shows no clinical symptoms.

Symptoms are fever; depression; decreased appetite; fatigue; rapid weight loss; swelling of legs, lower chest, and abdomen; irregular heartbeat; and sweating. Some animals may show no signs of being infected.

The only way to definitively diagnose EIA is by using the serum test known as the agar gel immunodiffusion (AGID) or Coggins test. An alternative (though possibly less accurate) test known as an ELISA (enzyme-linked immunosorbent assay) may be done when quick results are required.

There is no treatment and no cure for EIA, and currently no vaccine to prevent it. Federal and state laws require that all equines testing positive for EIA be identified by branding or tattooing and be strictly quarantined for life. Given these limited options, most owners opt to euthanize afflicted animals.

## Guttural Pouch Infection

The guttural pouch is an air-filled area in the donkey's throat. It is a large area that normally drains through a slit in the bottom of the nasal passage when the donkey lowers its head to eat. Occasionally, the slit will become clogged and the backed-up drainage will cause problems. This can happen following a bacterial (usually streptococcus) infection of the upper respiratory tract.

The symptoms include intermittent purulent nasal discharge, fever, labored breathing, depression, and decreased appetite. The veterinarian will usually makes the diagnois with an endoscopic examination. Flushing of the pouch and antibiotic therapy are the normal treatments. If the infection is chronic and severe, surgery is sometimes required to thoroughly drain the pouch.

## Hoof Abscesses

A hoof abscess is an infection within the foot of the donkey. Usually, there is a break somewhere in the sole of the hoof, which allows bacteria to get into the warm, moist environment of the inner hoof. The presence of bacteria leads to the formation of pus pockets.

The symptoms of a hoof abscess are often dramatic and unsettling to the donkey owner. Overnight, the donkey may become completely lame, unwilling to walk or even stand. Donkeys with abscesses spend an inordinate amount of time lying down. You may call your veterinarian, but often the farrier is the better choice for dealing with abscesses.

The vet or farrier may use an instrument known as a hoof tester, which looks like giant pincers. The tester is used to probe various areas of the sole in an effort to locate the abscess site. Often, donkeys don't react enough to the testers for them to be of much help. If the location is pinpointed, however, the vet or farrier may be able to pare away enough sole to reveal and drain the pus pocket. This will offer immediate relief to the donkey.

The area is then flushed with an antiseptic and usually soaked in a warm Epsom-salt solution to draw out and kill lingering bacteria. The clean, dry hoof may have a small plug of cotton soaked in 2 percent iodine solution inserted into the abscess prior to the whole hoof's being wrapped with duct tape. The hoof should be kept bandaged until the wound has ceased draining. You may want to give the donkey bute (an anti-inflammatory similar to aspirin) for pain. Occasionally, with an especially bad abscess, the vet will recommend a round of antibiotics.

The causes of hoof abscesses are many and varied, so prevention can be difficult. Wet conditions are thought to exacerbate them, so avoiding muddy, manure-filled stalls and paddocks will be beneficial. Proper hoof trimming can also help keep your donkey's feet healthy and free of abscesses.

## Pneumonia

Pneumonia is an inflammation of the lung. Common causes are bacteria, viruses, allergies, yeast, fungi, and inhalation of foreign material. Some predisposing factors that can lead to pneumonia are long trailer trips, viral respiratory disease, exposure to other diseased equines, choke, general anesthesia, and colic. Pneumonia is not generally contagious. Symptoms include fever, depression, rapid breathing, and difficulty breathing. The vet will form a diagnosis from a combination of physical exam (including listening to lung sounds with a stethoscope) and lab work (blood count). He or she may also do an ultrasound of the lungs or aspirate fluid from them for a culture. Antibiotic treatment is necessary to combat the disease.

To prevent pneumonia, avoid any stresses to the immune system of the donkey (long travel; overcrowding; exposure to excessive cold, rain, or snow). Keep the donkey vaccinated against respiratory viruses, and practice good barn hygiene.

## Rain Scald or Rain Rot

Rain scald or rot is one of the most common infectious skin diseases in equines. This is a bacterial infection caused by *Dermatophilus congolensis*. *Dermatophilus* persists in the environment and may be present in the soil. The donkey can carry the organism on its skin, but the organism may or may not cause an outbreak. Certain equines seem to be more predisposed to problems with it. Some factors that make an animal more prone to infection are immune-function related. Physical factors can include a thicker hair coat, thinner skin, habits that promote dirtier coats (such as frequent rolling), and insect bites. Donkeys definitely have the thicker coats and love nothing more than rolling in dirt, so it's no big surprise that they have frequent problems with this affliction. Rain rot outbreaks usually appear as crusty scabs or matted tufts of hair, which are easily scratched or lifted away from the skin. Although they often appear on the back, they can occur in other areas, as well.

Rain rot can go away by itself. However, if moist, warm conditions persist on the animal, the condition can progress into secondary bacterial infections and cause more problems. Rain rot is also contagious among other equines and can be spread via infected blankets and grooming tools. All equipment, including blankets, halters, and saddle pads, should be disinfected with a solution of 2 tablespoons of household bleach to 1 gallon of water. To prevent rain rot, clip off longer hair, if possible. Wash the animals with an antimicrobial and antibacterial shampoo. Keep the infected donkey as clean as possible, and separate it from any other donkeys exhibiting symptoms. In severe cases, penicillin injections may be required to aid healing.

## Recurrent Uveitis (Moonblindness)

Uveitis, also known as moonblindness, is a disease of the eye that can be caused by virus, bacteria, parasites, or injury. The condition requires prompt, aggressive treatment to save the animal's vision. In some cases, even treatment is not enough. Uveitis is an immune-mediated disease that results in repeated episodes, each one doing additional damage to the vision of the animal.

Symptoms of a uveitis episode include puffy, watery eyes, squinting and blinking, cloudiness or a blue/green tint to the eye, and tripping or running into objects.

Treatment usually consists of using atropine, which dilates the eye and reduces pain. A steroid and/or antibiotic ophthalmic ointment may be used as long as the eye is not ulcerated. (Symptoms of ulceration include squinting, watering eyes, and an animal's rubbing the eye against objects. A vet must diagnose ulceration.) Using steroids in an ulcerated eye can worsen the condition. Banamine (flunixin meglumine), bute (phenylbutazone), or aspirin are often administered to reduce inflammation. Because uveitis is tied in with immune function, many owners are experimenting with other methods of boosting the immune system. Protecting the donkey's eyes from sunlight and irritants by using a fly mask can help, as well.

# Rhinopneumonitis

Rhinopneumonitis is an equine herpes virus with symptoms similar to those of influenza. One noticeable difference is that coughing is not common with a rhino infection. Young animals are most vulnerable to the respiratory type of rhino, while pregnant jennies with the virus are at risk for late-term abortion.

Afflicted animals may have very mild symptoms, such as depression, moderate fever, "stocked-up" (swollen) legs, watery eyes, occasional nasal discharge, and rapid breathing. Unless the animals are further stressed, these symptoms will normally run their course in three to five days. The danger is if these animals come in contact with pregnant jennies during their infectious stage. If so, the jennies may begin spontaneously aborting their fetuses.

Vaccinations against rhinopneumonitis are available. The modified live virus vaccine is given annually, and there is a killed vaccine marketed to protect against abortion. The recommendation is that the vaccine be given during the fifth, seventh, and ninth months of a donkey's pregnancy.

# Sarcoid Tumors

Sarcoids are common, normally benign, tumors on donkeys and other equines. They are suspected to be caused by the bovine papillomavirus types 1 and 2. The tumors can be flat and patchlike or raised, knobby areas. Some are weepy, sore-looking lesions that may display a cauliflower-like appearance. They are typically found on the head and ears, but they may also appear around the tail area and on the legs.

The vet may do a biopsy to positively diagnose a sarcoid, but usually, a sarcoid is visually identifiable by location and characteristics of the lesion. Various treatments are used, including banding, surgical removal, freezing the tissue with liquid nitrogen, injection with an immune stimulant, and administering chemotherapy agents. No one treatment is suitable for all sarcoids; it must be determined on an individual basis. The chances of successfully eradicating the tumor are much greater when it is small. Don't wait to see if the tumor will disappear on its own. There is no method or vaccine available to prevent sarcoids.

# Strangles

Strangles is a very contagious equine infection caused by *Streptococcus equi* bacteria. It's most commonly seen in young animals, especially large groups of foals or yearlings. It is transmitted both directly and indirectly (buckets, pans, pasture environment, and contact with infected animals, as well as by flies).

Symptoms include a severe inflammation of the throat. Lymph nodes display extreme swelling and may rupture, producing a large amount of thick pus. The streptococcus organism is easily identified by a laboratory culture.

Argument exists over whether to treat strangles with antibiotics, as some people feel antibiotics interfere with the development of immunity and may actually prolong the disease. Generally, if the condition is caught early, treatment with penicillin is effective, but if it is more advanced, many vets advise owners to nurse the donkey through without use of antibiotics. Although vaccines are available, they are of questionable efficacy. They can reduce the rate of the disease in the face of an outbreak. Quarantine all new equines to your farm, and isolate any donkeys you suspect of having strangles.

## West Nile Virus (WNV)

WNV is a virus causing inflammation of the brain and/or the lining of the brain and spinal cord. It is not contagious between donkeys but is spread via mosquitoes. The virus infects and multiplies in birds. After mosquitoes bite infected birds, they are then capable of spreading the virus to humans, equines, and other mammals.

Symptoms of WNV can vary from mild to extreme. Typically, they include ataxia (incoordination), sleepiness, dullness, listless behavior, facial paralysis, and an inability to rise. Other signs can be fever, blindness, muscle trembling, and seizures. Diagnosis usually involves a positive blood-serum test combined with the display of clinical symptoms.

Most equines bitten by carrier mosquitoes do not develop clinical symptoms of the disease. Approximately one third of those that do become extremely sick are so severely afflicted that they must be euthanized. Treatment is limited to supportive therapy. Equines infected by the virus will develop antibodies, which can provide long-term immunity. There are vaccines available that help protect against West Nile virus.

# Glossary

**Afterbirth**—the placental material that is expelled following foaling

**Anaphylactic reaction**—an allergic reaction that can occur anywhere from a few minutes to a few hours following exposure to substances such as those in vaccinations

**Anthelmintic**—a deworming product

**Artificial insemination**—the placing by pipette of a jack's semen into the uterus of a jenny or mare by a person, in an attempt to impregnate her

**Bag**—the milk-filled udder

**Barren**—a female equine unable to get pregnant or to carry a foal to term

**Bite**—the alignment of an equine's front teeth. They should be as even as possible.

**Body clip**—using a pair of clippers (usually electric) to remove the coat of the donkey

**Bute**—term for the drug phenylbutazone. It is an analgesic similar to aspirin, used for equines.

**Castrate**—to surgically remove the testicles of the male donkey to render him incapable of impregnating a female and reduce aggression

**Chromosomes**—an organized piece of DNA found in the nucleus of cells

**Coggins test**—a blood test used to identify equine infectious anemia. Normally drawn on a yearly basis, sometimes more often depending on state laws. In most states, a negative test is needed to travel with, show, or sell donkeys.

**Colic**—a catchall term that basically describes a bellyache in the donkey. It can be extremely serious and needs assessment by a veterinarian.

**Colostrum**—the antibodies produced in the milk of the jenny immediately post-foaling. These antibodies are critical for the newborn to acquire and process in order to have immunity against disease and illness.

**Conception**—the time at which impregnation occurs

**Conformation**—a term describing the donkey's physical characteristics and how they are put together

**Covered**—mated

**CC (cubic centimeters)**—a unit of volume, often used in liquid equine medications. One cc equals one millimeter.

**Dam**—the mother of an animal

**Dorsal stripe**—the dark strip of hair extending from the withers, along the middle of the back, and down into the tail of most donkeys

**Drafty**—extremely heavy boned and heavy bodied

**Epinephrine**—a substance used to medically counteract an allergic reaction

**Estrone sulfate blood test**—a blood test used to detect pregnancy any time after 90 days postbreeding

**Estrus**—the period (usually lasting from three to eight days) in which the jenny is receptive to breeding and capable of becoming pregnant; also called *heat*

**Fallen crest**—the unsightly result of a donkey's neck becoming so fat and heavy that it falls over to the side. Usually, but not always, caused by obesity. Genetics play a role, as well. It is irreversible.

**Farrier**—a horseshoer; trims hooves and shoes donkeys, if needed

**Fertile**—capable of reproducing .

**Fetlock**—the "ankle" joint on an equine

**Fetus**—an unborn offspring far enough along that recognizable features of the mature animal are present

**Flank**—the area just in front of the haunch, or hindquarters, and just behind the belly of a donkey

**Fly predators**—tiny fly parasites that are purposely spread around manure piles and compost heaps. They will kill fly eggs before they hatch.

**Foal heat**—the first heat cycle a jenny experiences following foaling, usually nine to ten days later

**Foaling**—an equine's giving birth

**Forage**—grass-type fodder for donkeys

**Frog**—the V-shaped, triangular area on the bottom (sole) of the hoof

**Gaited**—usually means an equine that is capable of performing a smooth, non-jarring gait, such as a rack, a fox-trot, or a running walk

**Gelding**—a castrated male equine, or the act of castrating

**Gestation**—period a donkey is pregnant

**Granulated tissue**—new connective tissue that forms on the surfaces of wounds during the healing process

**Grazing**—to feed on growing grasses and herbage

**Halter**—a piece of equipment, usually nylon or leather, fitted on the equine to use to lead or to tie him

**Hand**—a unit of measurement for equines. One hand equals four inches. The measurement is taken from the ground to the top of the withers (shoulder blades).

**Hand breed**—breeding in a controlled situation in which both the jack and the jenny or mare are haltered and restrained; also known as *in-hand breeding*

**Harrow**—to drag a piece of equipment over the ground in an attempt to break up piles of manure and scatter them

**Hay**—grass or other plants cut and dried for use as livestock feed

**Heat**—see *Estrus*

**Herbivore**—an animal that feeds mainly on grasses and plants

**Herd**—a large group or collection of donkeys

**Hobbles**—a piece of equipment used to restrain the feet of an equine

**Hock**—the joint on the hind leg of the donkey that allows it to bend the leg

**Hybrid**—a cross between two different species

**Hybrid vigor**—the tendency for a cross-breed to have qualities superior to those of either parent

**In hand**—performing a breeding or showing event in which the donkeys are haltered and led

**Intramuscular (IM)**—method of injecting medicine by needle into the muscle

**Intravenous (IV)**—method of injecting medicine or fluids by needle into the vein

**Jack**—a male ass

**Jennet**—a female ass; also called a *jenny*

**Keratin tissue**—tough, insoluble protein substance found in hair, nails, horns, and hooves

**Lactating**—the act of producing milk and nursing a foal

**Legume**—type of hay from plants such as alfalfa, red clover, and lespedeza. Typically it is higher in protein than grass hay.

**Libido**—the sex drive

**Ligate**—to tie off blood vessels

**Lip twitch**—a device usually made from metal or a wood handle with a rope loop on the end. Applied to the muzzle of the donkey and twisted, it's used for subduing the animal for veterinary treatment, breeding, and other procedures.

**Lungworms**—worms that can affect all equines but normally can only reproduce in donkeys and mules. Donkeys and mules should be dewormed with ivermectin to avoid passing lungworms to horses.

**Maiden**—a jennet that has not yet produced a foal

**Mane**—the long hair that grows from the top of a donkey's neck

**Mare**—a female horse

**Meconium**—the first stool that the foal produces

**Medial patellar ligament**—a ligament that can be surgically split to treat exceptionally severe cases of upward fixation of the patella

**Metabolic rate**—the rate at which the body burns calories

**Muzzle**—a leather, nylon, or wire barrier that is strapped over a donkey's muzzle to keep him from biting or from overeating

**Oxytocin**—a drug that is commonly administered following foaling to cause uterine contractions that will help to expel a retained placenta, will assist in the "let-down" of milk, and will help to cleanse the uterus of any fluid or tissue left inside

**Overbite**—a malocclusion of the front teeth in which the upper teeth protrude over the lower ones

**Ovulation**—the release of a ripe egg from the ovaries

**Paddock**—a small, fenced enclosure for exercise or turnout of a donkey

**Palpation**—a method of pregnancy detection in which the veterinarian can feel an embryo through the rectal wall of the jennet/mare

**Papers**—usually meaning a donkey's registration certificate

**Pasture**—a good-size fenced enclosure usually used for grazing by the donkey

**Pasture breeding**—allowing a jack to run loose in a pasture with his jennies and breed them at will. Better conception rates result from pasture breeding, but sometimes injuries to both jack and jennies can occur.

**Pedigree**—a genealogical chart listing all of a donkey's known ancestors

**Placenta**—an organ formed in the donkey's uterus during pregnancy that envelops the fetus and provides oxygen and nutrients for it

**Quarantine**—keeping a donkey separate from other equines for a period of time, often used in cases of disease outbreak or when a new animal is brought to the farm

**Registered**—a purebred donkey whose information and pedigree are recorded in a specific breed registry organization

**Rotational grazing**—moving donkeys to different consecutive pastures to give grass time to regrow, limit parasite infestation, and improve the quality of forage

**Selenium**—a mineral found in the soil that is necessary for proper functioning of the donkey's immune system. Too much or too little can be harmful.

**Service**—the male donkey's act of breeding or mating

**Sire**—the father of an animal

**Sterile**—meaning incapable of reproducing

**Stifle**—a joint on the front of the donkey's hind leg; similar to the human knee

**Teaser**—a horse or pony stallion used to sexually excite a mare and convince her to allow herself to be bred by a jack

**Throatlatch**—the throat area on the donkey where the jaw and neck meet

**Transport company**—a company that may be professionally hired to haul animals that have been purchased, sold, or otherwise relocated

**Tubing**—passing a nasogastric tube through the nostril of the donkey during veterinary procedures

**Twinning**—producing twin offspring

**Ultrasound**—a procedure used to detect pregnancies and to rule out detrimental conditions such as twin conceptions

**Underbite**—a malocclusion of the front teeth in which the bottom teeth protrude further than the upper teeth

**Vaccination**—administration of vaccines to produce immunity to diseases

**Vestigial teat**—a nonfunctioning organ sometimes found on the sheath of the male donkey

**Vulva**—the external genital organ of the female donkey, located beneath the anus

**Wean**—to remove the foal from the dam and source of milk and accustom the foal to eating only solid feed

**White points**—white areas found around the eyes and on the muzzle, belly, and flank areas of most donkeys

**Wolf teeth**—vestigial (nonfunctioning) premolars usually found only on the upper jaw and commonly removed to prevent interference from a bit

# Resources

**D**onkey and mule information can be found through many sources, beginning with breed-specific associations, registries, and conservancies. Rescue and adoption organizations are also good places to get information and look at longears needing homes. Members of donkey and mule clubs and online lists will be happy to answer a newbie's questions. Of course, today the first place people look for information is on the Internet. Just bear in mind that a lot of online information has not been fact checked. Books and magazines, however, have been, as have videos, which also show you exactly how something is done. Contact information on various sources is listed below. Enjoy!

## Associations, Rescues, Clubs, and Conservancies

Associations and registries, rare breed conservancies, rescues, clubs, and online groups offer a plethora of information.

## Associations and Registries

These are the major North American donkey and mule registries and associations.

### American Council of Spotted Asses

914 Riske Lane
Wentzville, MO 63385
636-828-5955
www.spottedass.com/ACOSA_2/default
.aspx
Registry established to promote spotted asses.

### American Donkey and Mule Society

PO Box 1210
Lewisville, TX 75067
972-219-0781
www.lovelongears.com
Offers information on Miniature donkeys, standard donkeys, Mammoth Jackstock, mules, and zebra hybrids; provides services

for owners. Registers all types of longears. Membership includes a subscription to *The Brayer*.

### American Mammoth Jackstock Registry
PO Box 1723
Johnson City, TX 78636
830-330-0499
www.amjr.us
Registry that was established in 1888 for the American Mammoth Jackstock.

### American Miniature Mule Society
28565 N. Orion School Road
Canton, IL 61520
309-647-7162
www.miniaturemulesociety.com
Registers and brings together miniature mule breeders and enthusiasts.

### American Mule Association
260 Neilson Road
Reno, NV 89521
775-849-9437
www.americanmuleassociation.com
This association has been promoting and registering mules and donkeys since 1976.

### Canadian Donkey and Mule Association
25766 48th Avenue
Langley, BC V4W 1J2
604-857-4990
www.donkeyandmule.com
Canadian registry for donkeys and for mules. The association publishes *Canadian Donkey and Mule News*.

### International Miniature Donkey Registry
PO Box 982
Cripple Creek, CO 80813
719-689-2904
www.qis.net/~minidonk/imdr.htm

Incorporated in 1992 to register Miniature donkeys. It publishes *Miniature Donkey Talk* magazine.

### National Miniature Donkey Association
6450 Dewey Road
Rome, NY 13440
315-336-0154
www.nmdaasset.com
An association dedicated to the fostering and the improvement of the Miniature donkey. It publishes *The Asset*.

### North American Saddle Mule Association
PO Box 1008
Seneca, MO 64865
www.nasma.net
Registers and promotes saddle mules and donkeys.

## Rare Breed Conservancies
If you are interested in some of the rarer breeds of donkeys, here are a few organizations that you can contact for more information.

### American Livestock Breeds Conservancy
www.albc-usa.org
Nonprofit membership organization that is working to protect nearly 100 breeds of endangered, rare livestock, including Mammoth Jackstock. It's a pioneer in conserving historic breeds and genetic diversity.

### Asneria de J. M. Fernandez
http://asneriajmf.blogspot.com
This award-winning Spanish breeder of purebred Catalonian donkeys is dedicated to the restoration of this impressive breed. He owns one of the most prestigious Catalan farms.

## Association for the Protection of the Baudet de Poitou

(Association de Sauvegarde du Baudet)
www.baudet-du-poitou.fr/Sabaud.htm
An organization dedicated to the Poitou, an ancient French breed of ass.

## Murge Horse Information

(Cavallo delle Murge)
www.cavallodellemurge.it
A site about a rare breed of large Italian ass.

## National Association of Breeders of the Andalusian Donkey

(Asociacion Nacional de Criadores de la Raza Asnal Andaluza)
www.ancraa.org
A society dedicated to preserving the rare Andalusian donkey breed from Spain.

# Rescue and Adoption Organizations

If you are interested in adopting or would like to help out in another way, contact one of the organizations listed below.

## Crossroads Donkey Rescue

www.crossroadsdonkeyrescue.com
This Michigan group rescues abused and neglected donkeys in the Midwest.

## Donkey Sanctuary of Canada

www.thedonkeysanctuary.ca
This Canadian sanctuary rescues, rehabilitates, and protects abandoned, neglected, and abused donkeys, mules, and hinnies.

## The Donkey Sanctuary England

http://drupal.thedonkeysanctuary.org.uk
This sanctuary in Devon works to improve conditions for mules and donkeys.

## The Fund for Animals

http://fundforanimals.org
Founded by author and animal advocate Cleveland Amory, this nonprofit charity group works for the protection of all animals through sanctuary, rehabilitative, veterinary, and legislative efforts.

## Longhopes Donkey Shelter, Inc.

www.longhopes.org
This Colorado group rescues, rehabilitates, and adopts out standard donkeys.

## Nerja Donkey Sanctuary

www.nerjadonkeysanctuary.com
This small sanctuary in southern Spain rescues abandoned donkeys, mules, and horses. It also offers school programs and a riding program for the disabled.

## Peaceful Valley Donkey Rescue

www.donkeyrescue.org
This national organization seeks to improve the lives of American donkeys, through rescue, sanctuary, adoption, and education.

## Save Your Ass Long Ear Rescue

www.saveyourassrescue.org
This New England group rescues, rehabilitates, and rehomes donkeys and mules.

## Storybook Farm and Equine Rescue

www.storybook-farm.com
This private rescue group in Tunnel Hill, Georgia, is dedicated to the rescue of unwanted, neglected, and abused equines.

## Turning Pointe Donkey Rescue

www.turningpointedonkeyrescue.com
This Michigan rescue promotes the humane care and training of all donkeys.

## U.S. Bureau of Land Management

www.blm.gov/wo/st/en/prog/wild_horse_and_burro.html
In order to keep wild horse and burro herds in balance, the BLM removes a

certain percentage from the land each year and adopts out as many of them as possible.

## State and Regional Clubs

Want to find out more about longears? Attend a show? Visit a farm? Find an animal? Join a club and meet folks who will help.

**Arizona Mule and Donkey Association**
www.azmules.org

**California Donkey and Mule Association**
www.cadama.org

**Carolina Mule Association**
www.carolinamuleassociation.com

**Gulf Coast Donkey and Mule Association**
www.gcdma.org

**Illinois Draft Horse and Mule Association**
http://idhma.homestead.com

**Iowa Donkey and Mule Society**
www.idms.ws

**Kansas Draft Horse and Mule Association**
http://members.cox.net/kdhma

**Midstates Mule and Donkey Show Society**
http://midstatesmuleanddonkey.com

**Minnesota Donkey and Mule Club**
www.mndonkeyandmule.org

**Missouri Draft Horse and Mule Association**
www.missouridrafthorse.com

**Montana Draft Horse and Mule Association**
www.montanadrafthorsemule.com

**Oregon Mr. Longears Club**
www.oregonmrlongears.com

**Pikes Peak Long Ears Association**
www.pikespeaklongears.org

**Rio Grande Mule and Donkey Association**
www.rgmda.com

**Rocky Mountain Longears Association**
www.rockymountainlongears.com

**South Carolina Donkey and Mule Association**
www.myscdma.com

**Tennessee Donkey Association, Inc.**
www.tennessee-donkeys.com

**Southwestern Donkey and Mule Society**
www.southwesterndonkeymulesociety.com

**Virginia Draft Horse and Mule Association**
www.vdhma.org

**Western Pack Burro Association**
www.packburroracing.com

**Wisconsin Donkey and Mule Society, Inc.**
www.widonkeymule.org

## Online Discussion Groups

These online lists have been the source of great information, great friendships, and fabulous donkey purchases. It's fun to hook up with other "donkey nuts" and share!

## Donkeys
http://tech.groups.yahoo.com/group/donkeys
For the owners of donkeys and the owners of mules.

## Mammoth Donkeys
http://pets.groups.yahoo.com/group/mammothdonkeys
For breeders, owners, and admirers of Mammoths.

## Miniature Donkeys
http://pets.groups.yahoo.com/group/Minidonkey
For owners and other who are interested in minis.

## Mules
http://pets.groups.yahoo.com/group/MulesOnly
For owners and sellers of mules and mule goods.

# Books, Videos, and Periodicals
Some online sites are more reliable than others. Books and periodicals are fact checked, and they can be kept for easy reference. Videos offer you step-by-step demonstrations.

## Books and Videos
Here are a few favorite books and videos on the care and the training of donkeys and mules.

Gross, Bonnie. *Caring for Your Miniature Donkey*. 2nd ed. Miniature Donkey Talk, 1998.
A great resource for the Miniature donkey enthusiast.

Hodges, Meredith. *Donkey Training*. Lucky Three Ranch, 1999.
For beginners and veterans alike, this book covers the keys to training donkeys. Author Meredith Hodges also emphasizes the uniqueness of donkeys in the world of equines.

_____. *Training Mules and Donkeys: A Logical Approach to Longears*. Alpine, 1993.
A resistance training manual that covers simple psychology as well as techniques for training.

Hutchins, Betsy, and Paul Hutchins. *The Definitive Donkey: A Textbook of the Modern Ass*. Revised and edited by Leah Patton. American Donkey and Mule Society, 1999.
Another truly marvelous, must-have book for the donkey lover.

Svendsen, Elisabeth. *The Professional Handbook of the Donkey*. Whittet Books, 1998.
Published in Great Britain, this comprehensive handbook covers virtually all of the aspects of donkey health and donkey husbandry.

Ward, Crystal. *Donkey Training with Crystal Ward*. Available in both DVD and VHS formats from www.asspenranch.com.
Excellent information for new donkey owners. This video covers general donkey information, and provides tips for leading, ground driving, saddling, and riding.

Weaver, Sue. *The Donkey Companion*. Storey Publications, 2008.
A fabulous and extensive overview of information that you need to know about donkeys by one of the best writers in the business.

## Periodicals

These periodicals are good resources for tips on donkey health, training, showing, and breeding; they also list animals for sale.

### Asset
www.nmdaasset.com
315-336-0154
Miniature donkey publication put out by the National Miniature Donkey Association.

### The Brayer
www.lovelongears.com
972-219-0781
Published by the American Donkey and Mule Society, this magazine covers all longears.

### Miniature Donkey Talk
www.web-donkeys.com
719-689-2904
The International Miniature Donkey Registry publishes this magazine.

### Mules and More
www.mulesandmore.com
573-646-3934
A great, all-around magazine dedicated to mules and donkeys, emphasizing the former. There's an annual jack issue.

### Western Mule
www.westernmulemagazine.com
417-859-6853
This magazine mostly delves into the subject of mules, but it does have an occasional reference to donkeys. There's an annual jack issue.

## Sites for Buying Longears, Tack, and Other Supplies

Listed below are businesses and sites for buying and selling donkeys and mules, tack, veterinary supplies, and other equine supplies.

## Donkey, Mule, and General Equine Ad Sites

These sites have ads for buying and selling donkeys and mules. Sites with *horse* in the name also list donkeys and mules.

### DreamHorse.com
www.dreamhorse.com

### Eeebray.com
www.eeebray.com

### Equine.com
www.equine.com

### HorseTopia.com
www.horsetopia.com

### Gotdonkeys Breeders List
www.gotdonkeys.com

### LongearsMall.com
www.longearsmall.com

### Lucky Three Ranch
www.luckythreeranch.com

### The Mule Store
www.themulestore.com

### Muleville
www.muleville.com

## Tack

These businesses sell tack designed for donkeys and mules or fitting them well.

### Bob Marshall Saddles
www.sportssaddle.com
Treeless saddles are well regarded.

### Circle Y
www.circley.com
Their Flex-tree saddle is my favorite saddle for my donkeys.

## Queen Valley Mule Ranch

www.muleranch.com

Offering saddles designed and developed by mule owner and trainer Steve Edwards.

## Reed Tack

www.reedtack.com

Offers items made especially for longears.

## Tucker Trail Saddles

www.tuckersaddles.com

Selling saddles for the trail.

## Tuff Enuff

www.tuffenuff.org

Tack for trail riding and packing.

## Veterinary and Other Supplies

Vaccines, wormers, grooming supplies, and other equine equipment.

## Dover Saddlery

www.doversaddlery.com

800-406-8204

## Horse.com

(formerly Country Supply)

www.horse.com

800-637-6721

## Jeffers Equine

www.jeffersequine.com

800-533-3377

## KV Vet

www.kvvet.com

800-423-8211

## NASCO Farm & Ranch

www.enasco.com/farmandranch

800-558-9595

## State Line Tack

www.statelinetack.com

800-228-9208

## Valley Vet

www.valleyvet.com

800-419-9524

## Other Donkey Web Sites

Here are other sites offering training tips, health information, longear-themed gifts, and donkey genealogy resources.

## Blue Mountain Donkey Pages

http://bluemountainfarm.net/donkey page.htm

Created by former Mammoth breeder Jeanine Rachau, this is a series of articles on general care and training of donkeys.

## The Donkey Tree

www.donkeytree.com

Pat Scanlan's brainchild, this genealogical database of donkey pedigrees is now kept by Kristie Jorgenson of Sagebrush Ranch.

## Sagebrush Ranch
## Mammoth Jackstock

http://sbr.longearsmall.com

Webmaster Jorgenson, host of the Donkey Tree, provides pedigree service and LongearsMall.com classifieds; the site also offers many training tips.

## Shadow Ridge Donkey Links

www.shadowridgedonkeys.com/links.htm

A wonderful site, chock full of donkey information, with links to tons more.

## Team Donk

www.teamdonk.org

Written by Kristi Kingma of Idaho, this Web site has lots of information and instruction on donkey driving and care.

## WXICOF

www.wxicof.com

Great donkey and mule gift items, and many old and out-of-print donkey books.

# Photo Credits

The sources for the photographs and illustrations for this volume are listed below. Any images not credited here are copyright or courtesy of Anita Gallion.

# Index

## G

geldings
- about, 14, 41
- castration technique, 82, 94–95
- for guarding livestock, 29, 138

genetic defects
- malocclusion, 46, 98–99
- testicles fail to drop, 46, 98–99

gestation period, 20

grain, storing, 59

grass hay, 61–62

grazing muzzles, 56

Green, Robert, 14, 24, 25

guardians. *See* livestock guardians

gums, color of, 92

gut sounds, 92

guttural pouch disorders, 94, 150

## H

halters, removing, 68

halter training foals, 115

handling donkeys
- body clipping, 70
- disciplining miscreants, 75
- expert advice, 74
- feet, 72–73, 74
- foals, 115
- leading, 71–72
- saddling up, 76–77
- touching, 69, 71
- trailering, 47, 73, 75, 77
- training, 66–67, 69
- tying, 71

handling mules, 122–25, 127

hay, 61–62, 143

health care
- about, 79–80
- castration technique, 82, 94–95
- dental care, 82
- expert advice, 94–95
- farrier care, 72, 73, 82–83, 88, 140
- first aid kit, 85
- parasites, 83–87, 90
- vaccinations, 49, 80–82
- *See also* illnesses and diseases; veterinarians

health certificates, 49

heat cycle, 20

hinnies, 14, 123

hoof abscesses, 88, 150–51

horses
- breaking to lead with donkeys, 29
- donkeys as companions, 19, 29
- and lungworms, 84
- and mules, 124, 126–27
- taxonomy, 15

horses and donkeys compared
- about, 9
- castration technique, 82, 94–95
- colic symptoms, 91, 93
- disposition, 19–20, 71, 76–77
- feed requirements, 51, 53, 55, 61
- fencing, 57, 59
- flight instinct, 65
- mineral blocks, 62
- physical aspects, 17, 19
- reproductive tract, 107
- trailering, 77
- weaning, 115

horses and mules compared, 117, 119, 123, 124–25, 127

housing
- about, 51
- expert advice, 54–55
- fencing, 54, 55, 57, 59
- pasture, 53, 55, 63
- structure, 51–53

Houston Livestock Show and Rodeo, 44, 135

hybrid half-ass. *See* mules

hyperlipidemia, 56, 94, 95

## I

illnesses and diseases
- about, 87, 91, 92
- choke, 147
- colic, 85, 91, 93, 147–48
- Cushing's disease, 148
- EEE, 149
- EIA, 49, 149–50
- EPM, 148–49
- equine influenza, 148

**Anita Gallion** operates *Coyote Lane Farm* in west-central Illinois, where she is involved in breeding and showing American Mammoth Jackstock. Owner of the Yahoo Internet discussion list "Mammoth Donkeys" and a certified inspector for the American Donkey and Mule Society, Anita has written donkey-related articles, which have appeared in *Trail Rider* magazine and on the Illinois Horse Network. She may be reached at anita@saddledonkey.com.